T0229844

A CONCISE INTRODUCTION TO
PROGRAMMING IN
PYTHON
Second Edition

Chapman & Hall/CRC
Textbooks in Computing

Series Editors
John Impagliazzo
Andrew McGettrick

Ted Herman, A Functional Start to Computing with Python

Mark Johnson, A Concise Introduction to Data Structures Using Java

David D. Riley and Kenny A. Hunt, Computational Thinking for the Modern Problem Solver

Bill Manaris and Andrew R. Brown, Making Music with Computers: Creative Programming in Python

John S. Conery, Explorations in Computing: An Introduction to Computer Science and Python Programming

Jessen Havill, Discovering Computer Science: Interdisciplinary Problems, Principles, and Python Programming

Efrem G. Mallach, Information Systems: What Every Business Student Needs to Know

Iztok Fajfar, Start Programming Using HTML, CSS, and JavaScript

Mark C. Lewis and Lisa L. Lacher, Introduction to Programming and Problem-Solving Using Scala, Second Edition

Aharon Yadin, Computer Systems Architecture

Mark C. Lewis and Lisa L. Lacher, Object-Orientation, Abstraction, and Data Structures Using Scala, Second Edition

Uvais Qidwai, C.H. Chen, Digital Image Processing: An Algorithmic Approach with MATLAB

Mark J. Johnson, A Concise Introduction to Programming in Python, Second Edition

For more information about this series please visit:

https://www.crcpress.com/Chapman--HallCRC-Textbooks-in-Computing/book-series/CANDHTEXCOMSER?page=2&order=pubdate&size=12&view=list&status=published,forthcoming

Chapman & Hall/CRC
TEXTBOOKS IN COMPUTING

A CONCISE INTRODUCTION TO
PROGRAMMING IN
PYTHON
Second Edition

MARK J. JOHNSON

CRC Press
Taylor & Francis Group
Boca Raton London New York

CRC Press is an imprint of the
Taylor & Francis Group, an **informa** business

A CHAPMAN & HALL BOOK

CRC Press
Taylor & Francis Group
6000 Broken Sound Parkway NW, Suite 300
Boca Raton, FL 33487-2742

© 2018 by Taylor & Francis Group, LLC
CRC Press is an imprint of Taylor & Francis Group, an Informa business

No claim to original U.S. Government works

Printed on acid-free paper
Version Date: 20180314

International Standard Book Number-13: 978-1-1380-8258-8 (Paperback)

This book contains information obtained from authentic and highly regarded sources. Reasonable efforts have been made to publish reliable data and information, but the author and publisher cannot assume responsibility for the validity of all materials or the consequences of their use. The authors and publishers have attempted to trace the copyright holders of all material reproduced in this publication and apologize to copyright holders if permission to publish in this form has not been obtained. If any copyright material has not been acknowledged please write and let us know so we may rectify in any future reprint.

Except as permitted under U.S. Copyright Law, no part of this book may be reprinted, reproduced, transmitted, or utilized in any form by any electronic, mechanical, or other means, now known or hereafter invented, including photocopying, microfilming, and recording, or in any information storage or retrieval system, without written permission from the publishers.

For permission to photocopy or use material electronically from this work, please access www.copyright.com (http://www.copyright.com/) or contact the Copyright Clearance Center, Inc. (CCC), 222 Rosewood Drive, Danvers, MA 01923, 978-750-8400. CCC is a not-for-profit organization that provides licenses and registration for a variety of users. For organizations that have been granted a photocopy license by the CCC, a separate system of payment has been arranged.

Trademark Notice: Product or corporate names may be trademarks or registered trademarks, and are used only for identification and explanation without intent to infringe.

Library of Congress Cataloging-in-Publication Data

Names: Johnson, Mark J. (Mark James), 1961- author.
Title: A concise introduction to programming in Python / Mark J. Johnson.
Description: Second edition. | Boca Raton : Taylor & Francis, CRC Press,
2018. | Series: Chapman & Hall/CRC textbooks in computing | Includes index.
Identifiers: LCCN 2018011951 | ISBN 9781138082588 (pbk. : alk. paper)
Subjects: LCSH: Python (Computer program language) | Computer programming.
LC record available at https://lccn.loc.gov/2018011951

**Visit the Taylor & Francis Web site at
http://www.taylorandfrancis.com**

**and the CRC Press Web site at
http://www.crcpress.com**

Contents

CHAPTER 3 ■ Text   95

CHAPTER 4 ■ Images   145

CHAPTER 5 ■ Objects and Classes   167

List of Figures

List of Tables

Python Examples

Preface

Welcome!

This text provides an introduction to writing software in Python. No previous programming experience is necessary.

Work through the sections in order; projects and how-tos are optional. Each section begins with an example program illustrating one main new concept. For best results, follow these steps:

Read the section first to get the big picture.

Type and Run the example to see what it does.

Reread the section to dig into the details.

Write as many of the exercises as you can.

Review the section to make sure concepts are clear.

While it may not seem intuitive, you will learn by typing, both in terms of thoroughly reading and thinking about the programs, and through responding to error messages.

Writing software is a creative act: you make something new when you write a program. Exercises have prescribed ends but follow a developmental path, so that by working through this text, you will become able to create new programs for yourself.

Features

This second edition is thoroughly reorganized and rewritten, based on classroom experience, to incorporate:

- A spiral approach, starting with turtle graphics and then revisiting concepts in greater depth using numeric, textual, and image data

- Clear, concise explanations written for beginning students, emphasizing core principles

- A variety of accessible examples, focusing on key concepts

- Diagrams to help visualize new concepts

Designed for either classroom use or self-study, all example programs and solutions to odd-numbered exercises (except for projects) are available at

http://www.central.edu/go/conciseintro/

To Instructors

This text is designed for a first course in computer science and is suitable for majors and non-majors alike. In addition to the features outlined above, this second edition offers the following:

- Sections designed for approximately one class period each

- Early use of basic procedural constructs such as functions, selection, and repetition through turtle graphics

- A gradual development from procedural to object-oriented design

- Examples, exercises, and projects from diverse application domains, including finance, biology, image processing, and textual analysis

- New sections on recursion and exception handling, as well as an earlier introduction of lists, based on instructor feedback

- Use of the Pillow implementation of the Python image library PIL, compatible with Python 3

- A few brief How-To sections that introduce optional topics students may be interested in exploring

The text is written to be read, making it a good fit in flipped classrooms. Topics are introduced as needed, and the focus is always on what a beginning student needs to know rather than providing comprehensive documentation.

Feedback

Feel free to contact me at `johnsonm@central.edu`. I would appreciate hearing any comments, suggestions, or corrections you might have.

Acknowledgments

Many thanks to my colleagues Stephen Fyfe and Robert Franks for their conversations, suggestions, and continued support, and to Central College for the sabbatical that allowed time for a thorough revision. A special thanks goes to the students of COSC 110, whose questions and experiences have shaped this ongoing document. And thanks once again to Randi Cohen at Chapman & Hall/CRC Press for her sustained enthusiasm and support. Finally, I deeply appreciate Lyn's willingness to help me talk through difficulties. Someday, I will learn to have a normal conversation.

About the Author

Mark J. Johnson is professor of computer science and mathematics at Central College in Pella, Iowa, where he holds the Ruth and Marvin Denekas Endowed Chair in Science and Humanities. Mark is a graduate of the University of Wisconsin-Madison (Ph.D., mathematics) and St. Olaf College. He is the author of *A Concise Introduction to Data Structures Using Java*, also published by Chapman & Hall/CRC Press.

CHAPTER 1

Turtle Graphics

This chapter introduces many of the fundamental concepts of computing using examples from turtle graphics.

1.1 GETTING STARTED

Programmable software is what makes a computer a powerful tool. Each different program essentially "rewires" the computer to allow it to perform a different task. By following this text, you will learn basic principles of writing software in the Python programming language.

Python is a popular scripting language available as a free download from `www.python.org`. Follow the instructions given there to install the latest production version of Python 3 on your system. All examples in this text were written with Python 3.6.

The CPU and RAM

In order to write software, it will be helpful to imagine what happens inside the computer when a program runs. We begin with a rough picture and gradually fill in details along the way.

When a program is ready to run, it is loaded into RAM, usually from long-term storage such as a network drive or flash drive. **RAM** is an acronym for random access memory, which is the working **memory** of a computer. RAM is **volatile**, meaning that it requires electricity to maintain its contents.

Once a program is loaded into RAM, the **CPU**, or central processing unit, executes the instructions of the program, one at a time. Each CPU family has its own **instruction set**, and you might be surprised at how limited these instruction sets are. Most instructions boil down to one of a few simple types: load data, perform arithmetic, make comparisons, and store data. It is amazing that these small steps can be combined in so many different ways to build software that is incredibly diverse and complex.

Computer Languages

CPU instruction sets are also known as **machine languages**. The key point to remember about machine languages is that in order to be run by a CPU, a program *must* be written in the machine language of that CPU. Unfortunately, machine languages are not meant to be read or written by humans. They are really just specific sequences of bits in memory. (We will explain bits later if you are not sure what they are.)

Because of this, people usually write software in a **higher-level language**, in the sense of Table 1.1. This ordering is not meant to be precise, but, for example, most programmers would agree that C and C++ are closer to the machine than Python.

TABLE 1.1   Programming language hierarchy

Level	Language	Purposes
Higher	Python	Scripts
	Java	Applications
	C, C++	Applications, Systems
	Assembly Languages	Specialized Tasks
Lower	Machine Languages	

Compilation and Interpretation

Now if CPUs can only run programs written in their own machine language, how do we run programs written in Python, Java, or C++? The answer is that the programs are translated into machine language first.

There are two main types of translation: compilation and interpretation. When a program is **compiled**, it is completely translated into machine language to produce an executable file. C and C++ programs are usually compiled, and most applications you normally run have been compiled. In fact, many companies only distribute compiled executables: unless a project is **open source**, you do not have access to the uncompiled source code.

On the other hand, when a program is **interpreted**, it is translated "on-the-fly." No separate executable file is created. Instead, the translator program (the **interpreter**) translates your program so that the CPU can execute it. Python programs are usually interpreted.

The Python Interpreter

When you start Python, you are in immediate contact with a Python interpreter. If you provide it with valid Python, the interpreter will translate your code so that it can be executed by the CPU. The interpreter displays the version of Python it was written for, and then shows that it is ready for your input with a ">>>" prompt. The interpreter will translate and execute any legal Python code that is typed at this prompt. For example, if we enter the following statement:

```
>>> print("Hello!")
```

the interpreter will respond accordingly. Try it and see.

Remember that if you are not sure what something will do in Python, you can always try it out in the interpreter without having to write a complete program. Experiment—the interpreter will not mind.

A Python Program

That said, our focus will be on writing complete programs. Example 1.1 is an example of a Python program using turtle graphics.

```
1  # square.py
2  # Draw a square
3
4  from turtle import *
5
6  forward(100)
7  left(90)
8  forward(100)
9  left(90)
10 forward(100)
11 left(90)
12 forward(100)
13 left(90)
14
15 exitonclick()
```

Example 1.1 Draw a square.

Use your Python environment[1] to create a new file containing the code in Example 1.1 and save it as square.py. Python programs, also known as **scripts**, are stored in files ending in ".py." Run the program and observe the result. After the drawing finishes, click on the graphics window to close it. If that does not work, close the window manually or look for an option to restart the shell.

---

[1]If you are not sure what to use, start the Python **IDLE** application and choose "New Window" from the File menu. After typing the program and saving it, choose "Run Module" from the Run menu or press F5 to run the program.

This program illustrates many important Python concepts, which we will explore in the next few sections. It draws using a framework known as turtle graphics.

Python Turtle Graphics

Turtle graphics is a type of computer graphics that draws relative to the position of a virtual turtle on the screen. The turtle holds a pen, and if the pen is down when the turtle moves, then a line will be drawn. The turtle may also move with the pen up or initiate other types of drawing such as drawing a dot.

Table 1.2 describes the functions used in Example 1.1, along with a few others from the Python `turtle` module. Do not worry too much yet about what a "function" is—just start to get a feel for what these do.

TABLE 1.2   `turtle` module: basic functions

`forward(distance)` Move turtle forward *distance* units.
`backward(distance)` or `back(distance)` Move turtle backwards *distance* units.
`left(angle)` Turn turtle left *angle* degrees.
`right(angle)` Turn turtle right *angle* degrees.
`setheading(angle)` Rotate turtle to point in direction *angle*. 0 faces to the right.
`exitonclick()` Close turtle window when clicked.
`pendown()` Set turtle to draw when it moves.
`penup()` Set turtle to not draw when it moves.
`circle(radius, extent)` Move along circle of given *radius*. Optional *extent* specifies arc angle.
`dot(size)` Draw a dot at current location of optional *size*.

Syntax: Importing Python Modules

In order to use these Python turtle functions, they must be **imported** from the `turtle` module first, as in line 4 of Example 1.1. The **Python Standard**

Library consists of many **modules**, each of which adds a specific set of additional functionality to the basic language. To import a library module, use an **import** statement:

```
from module import *
```

The star at the end imports all available names from the given *module*. The star is helpful when you want to use many names from the same module, as we do in Example 1.1. Remember to import the **turtle** module at the beginning of all of your turtle graphics programs, and run the **import** statement first if you want to use turtle functions in the interpreter.

⟹ Caution: Do not name any of your program files **turtle.py** or you will not be able to import the **turtle** module.

Syntax: Comments

Comments begin with a pound sign **#** and signify that the rest of the line is to be ignored by the interpreter:

```
# Text that helps explain your program to others
```

The comments at the top of Example 1.1 give the name of the program and briefly describe its purpose. Comments may appear anywhere in a Python program and are meant for human readers rather than the interpreter, in order to explain some aspect of the code. Comments are used sparingly in this text, to reduce clutter as you read.

Using the Python Documentation

You can imagine that the **turtle** module must provide many other functions in addition to those listed above. You may also have questions about exactly how those functions work. The **Python Documentation** is online, extensive, and provides information like this and much more.

From the "Documentation" link at **www.python.org**, two links will be particularly useful: the Tutorial and the Library Reference.

> **Tutorial** provides informal descriptions of how most things work in Python. Use it when you start to learn a new topic.

> **Library Reference** is a good place to look up specific reference information. At the time of this writing, the complete list of functions in the **turtle** module is in Section 24.1 of the Library Reference for Python 3.6.

Be sure to use the documentation set that matches your version of Python.

Why Study Computer Science?

Finally, here are a few things to consider as we begin:

1. Software is everywhere, from tiny embedded systems to handheld mobile devices to massive warehouse data centers.

2. Computation is changing how other academic disciplines do their work, resulting in new interdisciplinary fields such as computational biology, computational linguistics, and digital humanities.

3. Programming develops your ability to solve problems. Because machine languages are so simplistic, you have to tell the computer *everything* it needs to do in order to solve a problem. Furthermore, running a program provides concrete feedback on whether or not your solution is correct.

4. Computer science develops your ability to understand systems. Software systems are among the most complicated artifacts ever created by humans, and learning to manage complexity in a program can help you learn to manage it in other areas.

5. Programming languages are tools for creation: they let you build cool things. There is nothing quite like getting an idea for a program and seeing it come to life. And then showing it to all your friends.

EXERCISES

1. At the start of a turtle program, in what direction is the turtle pointing?

2. At the start of a turtle program, is the turtle's pen up or down?

3. Compare the results of running these two sequences of steps:

   (a) `forward(100)`   (b) `left(90)`
        `left(90)`   `forward(100)`

4. Compare the results of running these two sequences of steps:

   (a) `backward(100)`   (b) `right(90)`
        `right(90)`   `backward(100)`

5. Determine the center of the circle drawn by this program:

   `circle(100)`

6. Determine the center of the circle drawn by this program:

   `left(90)`
   `circle(100)`

7. Sketch the result of these turtle steps:

   ```
   left(45)
   forward(141)
   left(135)
   forward(100)
   left(90)
   forward(100)
   ```

8. Sketch the result of these turtle steps:

   ```
   forward(100)
   right(90)
   circle(50)
   right(90)
   forward(200)
   right(90)
   circle(50)
   ```

9. Write a program to draw a rectangle 200 units wide and 100 units tall.

10. Write a program to draw a rectangle 100 units wide and 200 units tall.

11. Write a program to draw an equilateral triangle with side length 100.

12. Write a program to draw a right triangle of any size.

13. Write a program to draw a circle centered at (100, 50) with radius 100. Put a dot at the center of the circle.

14. Write a program to put a circle inside the square of Example 1.1, so that the circle just touches the edges of the square.

15. Experiment moving the turtle to determine the coordinate system used in the graphics window. Where is the origin? What are the coordinates of the lower-left and upper-right corners? What are the width and height of the window?

16. Describe in your own words where the circle is relative to the turtle that the circle() function draws.

17. Write a program to draw something of your choice.

18. Use the Python documentation to learn about a turtle function not listed in Table 1.2. Use this function in a program to draw something.

1.2   CALLING FUNCTIONS

The turtle functions in Section 1.1 require you to know what angles to turn and what distances to move. Example 1.2 demonstrates another way to direct the turtle. The new functions used in this example are listed in Table 1.3. Try them out to draw your own shapes. As you do, begin paying attention to the terminology of Python programs and functions, as described in the rest of this section.

```
1   # bowtie.py
2   # Draw a bowtie
3
4   from turtle import *
5
6   pensize(7)
7   penup()
8   goto(-200, -100)
9   pendown()
10  fillcolor("red")
11  begin_fill()
12  goto(-200, 100)
13  goto(200, -100)
14  goto(200, 100)
15  goto(-200, -100)
16  end_fill()
17
18  exitonclick()
```

Example 1.2   Draw a bowtie.

Program Syntax and Semantics

In order to compile or be interpreted, a program must be written with the correct **syntax**, meaning the precise form dictated by the language designers. If a program has the correct syntax, then its **semantics** are defined by what the program does when it runs. With spoken languages, semantics refers to the meaning of statements; a program's meaning is given by what it does.

Program errors fall into two broad categories based on these terms:

> **Syntax errors** are mistakes in the form of the program. They will be reported by the compiler or interpreter, usually at the first point where the translation failed.

> **Semantic errors** are present when a program runs but does not do what its author intended. These cannot be reported by the

TABLE 1.3 **turtle** module: goto, color, fill, hide

goto(x, y)   Move turtle to location (x, y).
pensize(width)   Set thickness of line drawn by pen to *width* or return current value.
pencolor(color)   Set pen color to *color* or return current value.
fillcolor(color)   Set fill color to *color* or return current value.
color(color)   Set both pen and fill color to *color* or return current values.
begin_fill()   Begin a shape to be filled.
end_fill()   End the shape to be filled.
turtlesize(factor)   Stretch turtle by *factor*.
showturtle()   Start displaying turtle.
hideturtle()   Stop displaying turtle.

interpreter, because the interpreter has no way of knowing what a program is supposed to do.

By their nature, semantic errors are much more varied and difficult to locate and correct—there are many ways to write programs that do not do what we want!

Program Flow of Control

A program is a sequence of **statements**, which are individual commands executed one after another by the Python interpreter. As noted already, every statement must have the correct syntax in order to be translated.

Although it may sound obvious, the fact that statements execute one after the other—in the order they are given in the program—is essential to grasp. Visualize this **flow of control** as a downward stream of steps, as shown in Fig. 1.1. You will learn different ways to alter this flow, but for now, try to internalize the idea that programs work step-by-step, according to the order in which you specify its statements.

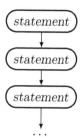

Figure 1.1 Normal flow of control.

Python Functions

A **function** in Python is a bundle of code designed to perform a specific task. Functions in Python are either built-in, imported from a module, or written as part of a program. **Built-in functions**[2] are always available and do not need to be imported. In contrast, the turtle functions in Tables 1.2 and 1.3 all come from the `turtle` library module, so they need to be imported. Each turtle function has a well-defined purpose captured by its **name**. Function names are important, since if a library provides a function named `f()`, you will have a hard time knowing or remembering what it does.

Function Parameters

Functions communicate with each other through optional parameters and return values. **Parameters** are data items that the function takes as input to use during its execution. We will consider return values later, beginning in Section 1.5.

Example: A function with no parameters

> The simplest functions require no communication and so have no parameters. For example, the `penup()` function causes the turtle to stop drawing as it moves. This function requires no additional information in order to know what to do.

Example: A function with two parameters

> Functions have parameters if they need additional information to accomplish their task. For example, the `goto()` function needs the x and y coordinates of the point to move to; visualize this as in Fig. 1.2. Values for parameters will be specified when the function is called (see below).

---

[2]See the index for a list of the built-in functions used in this text.

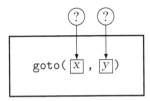

Figure 1.2 Function parameters.

If you imagine a different function `goto_origin()` with no parameters that always goes to the origin, you can see how parameters help make functions *general-purpose* as opposed to specific.

Syntax: Calling Functions

Every function has the capacity to perform a task, but it only performs that task when it is called. A function **call** requests execution of the function with particular **arguments** passed as the values for its parameters (if any).

The syntax of a function call is:

```
name(argument1, argument2, ...)
```

When this expression appears in a program statement that is being executed, the function called *name* executes, using the argument values inside parentheses as the values of its parameters.

Example: `goto(-200, 100)` called on line 12 of Example 1.2

> In this case, the values −200 and 100 are the arguments being sent to the `goto()` function for the parameters x and y; see Fig. 1.3. This call causes the turtle to move to location $(-200, 100)$.

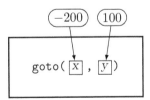

Figure 1.3 Function arguments.

In general, a function has one set of parameters, but it can be called with any number of different sets of arguments. In the case of `goto()`, its parameters are always x and y, but it can be called with whatever arguments you need, as on line 8 and lines 12–15. This flexibility is an important part of the power of functions in programming languages.

EXERCISES

1. Use Tables 1.2 and 1.3 to list the turtle functions we have seen so far with no parameters.

2. Use Tables 1.2 and 1.3 to list the turtle functions we have seen so far with one parameter. Give the name of the parameter in each case.

3. Draw a diagram like Fig. 1.3 for the function call on line 8 of Example 1.2.

4. Draw a diagram like Fig. 1.3 for the function call on line 14 of Example 1.2.

5. Classify each of these errors as syntactic or semantic:

    (a) Writing `beginfill()` in a turtle program

    (b) Forgetting to call `penup()` to stop drawing

    (c) Mistyping `left()` as `lft()`

6. Classify each of these errors as syntactic or semantic:

    (a) Forgetting to change the pen size

    (b) Forgetting to include a necessary **import** statement

    (c) Mistyping `left()` as `right()`

7. Sketch the result of these turtle steps:

```
left(90)
forward(200)
goto(70, 170)
goto(0, 140)
```

8. Sketch the result of these turtle steps:

```
goto(-100, 0)
goto(100, 0)
penup()
goto(0, -100)
pendown()
circle(100)
```

9. Write a program to draw the same square as Example 1.1 using `goto()` instead of `forward()` and `left()`.

10. Write a program to draw the triangle having vertices $(-100, 300)$, $(150, 50)$, and $(10, -200)$ with a thick border and blue interior.

11. Write a program to draw a square of width 100 centered at $(-50, 150)$. Put a dot at the center of the square.

12. Write a program to draw a circle centered at (100, 50) with radius 100. Put a dot at the center of the circle.

13. Write a program to draw a silo: a vertical rectangle with a semicircle on top. Use the optional *extent* parameter of the `circle()` function.

14. Write a program to draw an ice cream cone: a tall triangle with a semicircle on top. Use the optional *extent* parameter of the `circle()` function.

15. Write a program to draw the message "Hi" on the screen without using the `write()` function. Hide the turtle so that only the message is visible.

16. Write a program to draw the message "Go" on the screen without using the `write()` function. Hide the turtle at the end.

17. Write a program to re-create the cover image (a droodle by Roger Price) of Frank Zappa's *Ship Arriving Too Late to Save a Drowning Witch*.

18. Write a program to create an image in the style of Dutch painter Piet Mondrian's *Composition with Red, Blue, and Yellow* from 1930.

1.3 WRITING FUNCTIONS

In some applications, it might be helpful to draw a circle by specifying its center and radius. But you may have noticed that the turtle needs to start at the *bottom* of the circle and then call the `circle()` function—as long as the turtle is pointing to the right, as in Fig. 1.4.

Figure 1.4 Circle drawn by turtle graphics.

Thus, to draw a circle with center at (x, y) and radius r, the turtle needs to move to the bottom of where the circle will be (without drawing), point to the right, and then draw. Example 1.3 shows how to bundle these steps together into a new `circle_at()` function that can be used any time we are given a center and radius.

Notice how lines 14–20 use the `circle_at()` function to draw three circles. Using functions you write yourself is no different from using any of the library or built-in functions. The part that is new is defining the `circle_at()` function.

```
1  # circle_at.py
2  # Use a function to draw circles
3
4  from turtle import *
5
6  def circle_at(x, y, r):
7      """Draw circle with center (x, y) radius r."""
8      penup()
9      goto(x, y - r)
10     pendown()
11     setheading(0)
12     circle(r)
13
14 circle_at(-200, 0, 20)
15 begin_fill()
16 circle_at(0, 0, 100)
17 end_fill()
18 circle_at(200, 0, 20)
19 hideturtle()
20 exitonclick()
```

Example 1.3 Circle function.

Syntax: Defining Functions

A function **definition** specifies the code that will be executed when the function is called. Remember from the last section that when we call a function, we ask it to do something—it is the function's definition that describes exactly what will be done.

The syntax of a function definition is:

```
def name(parameter1, parameter2, ...):
    statement
    statement
    statement
    ...
```

Function definitions have two parts:

> **Header** The first line of a function definition, known as the function **header**, begins with **def**, followed by the *name* of the function and an optional list of parameters inside parentheses. The parentheses are always required, but remember that some

functions do not need parameters. The header ends with a colon, which introduces the body.

Body The body of the function is made up of all the statements that are indented underneath the header. Code that is indented underneath a line ending in a colon is called a **block** or **suite**. The block ends at the first line that is *not* indented. Blank lines are allowed inside a block.

The body of a function is the code that will be executed when the function is called. In Example 1.3, lines 8–12 are the body of the `circle_at()` function. (We will consider line 7 shortly.)

Function definitions usually appear near the top of Python programs, after any **import** statements, but before any directly executable code. Lines 14–20 of Example 1.3 are directly executable, like all of the code in our previous examples.

Docstring Comments

Docstring comments describe functions to other programmers. Docstrings are enclosed in triple quotes (three double-quote " characters), and may consist of any number of lines. The first line should be a brief description of the function's purpose; for example, line 7 of Example 1.3 documents the `circle_at()` function. Like other comments, docstrings are not executed.

Naming Functions and Parameters

As you write your own functions, think carefully about their names and the names of their parameters. There are few hard-and-fast rules, but try to choose names that are meaningful in the context they will be used. Be careful of very short names: `f(x)` is almost always too vague, but in the context of `circle_at()`, the meanings of the names x, y, and r are reasonably clear.

The only mandatory requirement is that function and parameter names must be legal Python **identifiers**. The actual rules for identifiers are contained in the Python Language Reference, but for our purposes, identifiers consist of upper and lower alphabetic characters (including non-English characters), the underscore (_), and, except for the first character, the digits 0–9. There is a small set of reserved Python **keywords** that may not be used as names. These keywords appear in bold in program examples, such as **from**, **import**, and **def** in Example 1.3, and are listed in the index.

Abstraction: Functions Help You Think

Look again at lines 14–20 of Example 1.3. When you use the `circle_at()` function, you no longer worry about how to get a circle centered at (x, y)—you just call the function. This is similar to the way we called library functions: we didn't worry about how `forward()` worked, we just called it to move the turtle forward. In other words, by writing new functions, we are able to *ignore details* and think at a higher level, in this case, by thinking in terms of drawing a circle at a location. This technique is called **abstraction**. Abstractions allow you to ignore lower-level details in order to think in terms of higher-level concepts.

Calling a Function Changes the Flow of Control

The flow of control changes in order to enable this abstraction. A function call creates a diversion in the flow of control to execute the body of the function, as shown in Fig. 1.5. This diversion makes the abstraction work: you can think of the function call as a single step, without worrying about the detail in the function body.

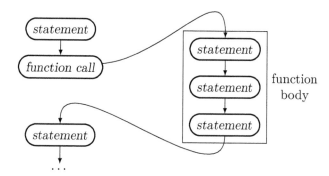

Figure 1.5 Function call flow of control.

EXERCISES

1. Draw a diagram like Fig. 1.3 showing arguments and parameters for each call to `circle_at()` in Example 1.3.

2. Draw a diagram like Fig. 1.5 for Example 1.3, showing the flow of control of the entire program. Only expand calls to the `circle_at()` function. In other words, consider calls to library functions as single steps, even though they are not.

3. Write a program using the `circle_at()` function to draw a bullseye. Type the `circle_at()` function in your program in order to use it.

4. Use the `circle_at()` function to write a `bullseye_at(x, y)` function that draws a bullseye centered at (x, y) with extra-thick lines. Use it to draw several targets.

5. Write a program using the `circle_at()` function to draw three mutually overlapping rings:

6. Write a program using the `circle_at()` function to draw the Olympic rings in color. Do not worry about the interlocking.

7. Write a `square_at(x, y, width)` function to draw the square centered at (x, y) with the specified *width*. Draw using `forward()` and `left()`, as in Example 1.1. Use your function to draw squares where there are circles in Example 1.3.

8. Write a `square_at(x, y, width)` function to draw the square centered at (x, y) with the specified *width*. Draw the square using `goto()`. Use your function to draw squares where there are circles in Example 1.3.

9. Write a `rectangle_at(x, y, width, height)` function to draw the rectangle centered at (x, y) with the specified *width* and *height*. Draw using `forward()` and `left()`. Use your function to draw three rectangles.

10. Write a `rectangle_at(x, y, width, height)` function to draw the rectangle centered at (x, y) with the specified *width* and *height*. Draw using `goto()`. Use your function to draw three rectangles.

11. (Requires either Exercise 9 or 10.) Use a `rectangle_at()` function to write a `square_at()` function. Use your functions to draw some squares and rectangles.

12. Write a `square(width)` function to draw a square so that the lower-left corner is put where the turtle is located and the square is oriented in the direction of the turtle. Use your function to draw some squares.

13. Write a `rectangle(width, height)` function to draw a rectangle so that the lower-left corner is put where the turtle is located and the rectangle is oriented in the direction of the turtle. Use your function to draw some rectangles.

14. Write a `triangle(x0, y0, x1, y1, x2, y2)` function to draw the triangle with vertices $(x0, y0)$, $(x1, y1)$, and $(x2, y2)$. Use your function to draw some triangles.

15. Write a `silo(width, height)` function to draw a silo shape: a vertical rectangle with a semicircle on top. Draw so that the lower-left corner is wherever the turtle is located.

16. Write two functions, `drawH()` and `drawi()` to draw the letters "H" and "i," respectively, and then use these functions to draw the message "Hi."

1.4 REPETITION

Much of the power of computing comes from its ability to automate repetitive tasks. Look again at the code to draw a square from Example 1.1. It can be summarized as:

```
repeat 4 times:
    forward(100)
    left(90)
```

Drawing an octagon is similar:

```
repeat 8 times:
    forward(100)
    left(45)
```

There is a pattern here, and Example 1.4 shows how to express the pattern in Python.

```
 1  # polygon.py
 2  # Draw regular polygons
 3
 4  from turtle import *
 5
 6  def polygon(n, length):
 7      """Draw n-sided polygon with given side length."""
 8      for _ in range(n):
 9          forward(length)
10          left(360/n)
11
12  def main():
13      """Draw polygons with 3-9 sides."""
14      for n in range(3, 10):
15          polygon(n, 80)
16      exitonclick()
17
18  main()
```

Example 1.4   Draw polygons.

Main Function

Example 1.4 is written with a **main()** function, instead of simply putting the executable statements after other function definitions. Functions like **main()** that are intended to drive the execution of a program are sometimes known as

drivers. This is a common way of writing Python programs, so we will follow it for the rest of this text.

⟶ Note: The call to `main()` on line 18 is important—without it, `main()` would be defined but never run.

Syntax: `for` Loops

A **for** loop is one of the main tools in Python for repeating steps in a program. They are most useful when you know ahead of time how many times the loop should run. The syntax of a **for** loop is:

```
for variable in sequence:
    body
```

A **for** loop works like this:

- The *body* is executed once for each item in the *sequence*.

- The *variable* is a name for the item from the sequence, which can be used in the *body*.

The *body* of a **for** loop is like the body of a function: it is the indented block that follows the colon at the end of the **for** statement. Remember that a block ends when the indentation ends.

There is a lot to learn about loops, variables, and sequences that will come later. For now, our sequences will come from the **range()** function, and variable names may be any legal Python identifier (see page 15).

Ranges

The built-in **range()** function is often used to create sequences in Python **for** loops. The **range()** function may be called in three different ways, described in Table 1.4. All parameters must be integers.

TABLE 1.4 The **range()** function

range(stop**)**   Generate sequence 0, 1, 2, . . . , $stop - 1$
range(start, stop**)**   Generate sequence $start$, $start + 1$, $start + 2$, . . . , $stop - 1$
range(start, stop, step**)**   Generate sequence $start$, $start + step$, $start + 2 \cdot step$, . . . , ending at the last term before reaching $stop$

Example: range(10) generates 0, 1, 2, 3, 4, 5, 6, 7, 8, 9.

> If you do not specify a *start* value, range sequences start at 0 and stop *before* reaching *stop*, in this case, at 9. It is important to see that this sequence has 10 items in it. That means a loop over **range(10)** runs 10 times, and, in general, that a loop over **range(n)** runs n times.

Example: range(2, 7) generates 2, 3, 4, 5, 6.

> This range starts at 2 and stops before reaching 7. When using two parameters like this, it is helpful to recognize that $7 - 2 = 5$ items are generated.

Example: range(1, 10, 2) generates 1, 3, 5, 7, 9.

> The third parameter specifies a step size, and the items stop before reaching the *stop* value.

Example: range(5, 2, -1) generates 5, 4, 3.

> The step size may be negative. Ranges always end at the last term before reaching the *stop* value, whether they are increasing or decreasing.

The **range()** function is not intuitive at first. Ranges start at 0 unless you specify otherwise, and they end *before* the stopping point.

⟶ Note: The interpreter will not directly show you the values in a **range()**. To see them, use the built-in **list()** function, like this:

```
list(range(5, 2, -1))
```

which returns [5, 4, 3].

Loop Variables

Compare the two loops in Example 1.4, beginning with the second one.

Example: for n in range(3, 10) on line 14

> This loop inside **main()** uses the value of **n** to draw different polygons. It is equivalent to writing out the body with each different value of **n**:
>
> ```
> polygon(3, 80)
> polygon(4, 80)
> ...
> ```

Example: for _ in range(n) on line 8

> This loop simply needs to run its body **n** times, so an **underscore** is used as the variable name to show that the individual items from the range will be ignored. (In other words, the values $0, 1, \ldots, n-1$ are never used in the loop body.) Variables like this whose values are never used are called **dummy variables**.

for Loop Flow of Control

The normal flow of control is that statements execute one after the other, in the order they are written. A **for** loop changes this flow to repeat the body for each item in the given sequence, as shown in Fig. 1.6. After the last item in the sequence, the loop finishes and control continues as usual.

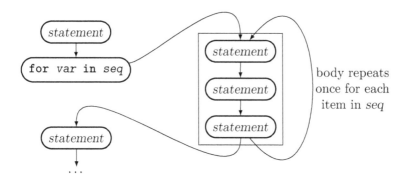

Figure 1.6 **for** loop flow of control.

EXERCISES

1. Identify the body of the **for** loop beginning at line 8 of Example 1.4.

2. Identify the body of the **for** loop beginning at line 14 of Example 1.4.

3. List the items generated by **range(n)** from line 8 of Example 1.4. (These values are ignored in the body of the loop.)

4. List the items generated by **range(3, 10)** from line 14 of Example 1.4.

5. Determine the sequence generated by each of these range expressions:

   (a) **range(10)**   (c) **range(10, 5)**

   (b) **range(5, 10)**   (d) **range(2, 11, 3)**

6. Determine the sequence generated by each of these range expressions:

    (a) `range(10, 0, -1)`   (c) `range(0, 10, -1)`

    (b) `range(10, 0, -2)`   (d) `range(0, 1, 0.1)`

7. Determine a range expression to produce each of these sequences:

    (a) `0, 1, 2, 3`   (c) `1, 3, 5, 7, 9, 11`

    (b) `3, 2, 1, 0`   (d) `4, 7, 10, 13, 16, 19, 22`

8. Determine a range expression to produce each of these sequences:

    (a) `10, 27, 44, 61, ..., 197`   (c) `2, 7, 12, 17, ..., 2037`

    (b) `1000, 900, 800, ..., 100`   (d) `200, 198, 196, ..., 0`

9. Write a program with a loop to draw circles of radius $r = 10, 20, 30, \ldots, 80$. Use the `circle()` function, so they all begin drawing from the same point.

10. Use the `circle_at()` function from Section 1.3 to write a program that draws circles centered at the origin with radius $r = 10, 20, \ldots, 80$. Use a loop.

11. Use the `circle_at()` function to write a program that draws a line of seven circles with radius 50 at locations $(-300, 0)$, $(-200, 0)$, $(-100, 0)$, $\ldots (300, 0)$. Use a loop.

12. Use the `circle_at()` function to write a program that draws a line of seven circles with radius 70.71 at locations $(-300, -300)$, $(-200, -200)$, $(-100, -100)$, $\ldots (300, 300)$.

13. Write a loop for `y` inside of a loop for `x` (called nested loops) to draw a grid of circles with radius 50 spaced 100 units apart, with centers $(-200, -200)$ in the lower-left corner to $(200, 200)$ in the upper-right corner. Describe the order in which the circles are drawn.

14. Use the `circle_at()` function to write a `bullseye_at(x, y, radius)` function that draws a bullseye centered at (x, y) of the given radius with extra-thick lines. Draw more rings when the radius is larger—as many as will fit. Use your function to draw three targets of different sizes.

15. Write a program to repeat these steps 36 times: draw a square, and then turn left 10 degrees. Use the `polygon()` function to draw the squares.

16. Write a program to repeat these steps for $n = 5, 10, 15, \ldots, 200$: move forward n steps, and then turn left 90 degrees. Try to predict the output.

17. Write a program to draw a spiral shape by having the turtle move forward an increasing amount while turning left a decreasing amount.

18. Use the `polygon()` function to draw a circle.

19. Write a `mycircle(radius)` function that uses the `polygon()` function to mimic the library `circle()` function. Use your function in a loop to draw several circles. Hint: choose a value for n, and then use side length `6.28 * r / n`.

1.5   CHECKING CONDITIONS

As we have seen, programs execute statements one after the other until a statement changes the flow of control. **Conditional statements** allow a program to execute different code depending on what happens as the program runs. The name comes from the fact that true-false **conditions** are used to decide which code to execute. Conditional statements are also known as **selection** statements because testing a condition can be thought of as selecting an option. This flexibility to make choices while a program runs is another important key to the power of computation.

```
1  # bounce.py
2  # Bounce the turtle.
3
4  from turtle import *
5
6  def move(distance):
7      """Move forward, reversing direction at right side."""
8      forward(distance)
9      if xcor() > 320:
10         setheading(180)
11
12 def main():
13     shape("circle")
14     penup()
15     speed(0)
16     for _ in range(100):
17         move(10)
18     exitonclick()
19
20 main()
```

Example 1.5   Bouncing turtle.

Example 1.5 uses repetition along with Python's conditional statement to animate a bouncing ball. The **if** statement handles bouncing by reversing the turtle's direction *if* it approaches the right-hand edge. New turtle functions used in this program are given in Table 1.5.

TABLE 1.5   `turtle` module: shape, speed, location, heading

`shape(name)`   Change turtle's shape to *name*.
`speed(value)`   Set turtle's speed to *value* between 1 (slow) to 10 (fastest), with 0 being "instantaneous."
`xcor()`   Return current *x*-coordinate of turtle.
`ycor()`   Return current *y*-coordinate of turtle.
`position()`   Return current coordinates of turtle as (x, y) pair.
`heading()`   Return current heading of turtle in degrees.
`towards(x, y)`   Return heading toward (x, y) from current location.
`distance(x, y)`   Return distance to (x, y) from current location.

Syntax: `if` Statements

Python's conditional statement is the `if` statement, which can be written in several different forms. The simplest is:

```
if condition:
    body
```

In this form, *condition* is evaluated, and if it is true, then *body* is executed. If the condition is false, then *body* is not executed. We will see other forms of the `if` statement in Section 1.7.

`if` Statement Flow of Control

An `if` statement changes the normal flow of control to allow executing a group of statements only if some condition holds. Figure 1.7 shows one way to visualize this, where the body of the `if` statement is a side route taken only when the condition is true.

Boolean Expressions

Conditions in Python are written as **boolean expressions**, which are expressions that evaluate to either `True` or `False`. In Python, `True` and `False` are

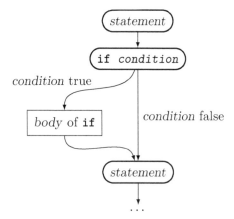

Figure 1.7 **if** statement flow of control.

always capitalized and are keywords. Table 1.6 lists Python's **comparison** operators, which all return boolean values.

TABLE 1.6 Comparisons

x == y
True if x and y are equal; otherwise, **False**.
x != y
True if x and y are not equal; otherwise, **False**.
x < y
True if x is less than y; otherwise, **False**.
x > y
True if x is greater than y; otherwise, **False**.
x <= y
True if x is less than or equal to y; otherwise, **False**.
x >= y
True if x is greater than or equal to y; otherwise, **False**.

⟹ Caution: Use == to test for equality in Python, not =.

Using Functions with Return Values

In addition to whatever work they do, functions have the option of **returning** a value to the caller. When calling functions, you want to be aware of whether or not there is a return value that you are expected to use. Example 1.5 illustrates the difference. First consider the call to **forward()** in line 8:

```
forward(distance)
```

The **forward()** function does not have a return value. When it is called, the turtle should move forward the given **distance**, and that is all. Functions without return values are sometimes referred to as **procedures** to emphasize the difference.

Compare this to calling **xcor()** on line 9:

```
if xcor() > 320:
```

Here, the **xcor()** function returns the current x-coordinate of the turtle. Notice how the return value is used: the *call* is treated as if it is the *value* being returned. Suppose the current x-coordinate is 100. Then we can imagine the **if** statement as in Fig. 1.8, where the call to **xcor()** is replaced by its return value of 100.

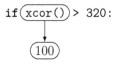

Figure 1.8 Function call replaced by its value.

You probably took this for granted in reading Example 1.5. But it is important to become aware of using return values as you write your own programs.

EXERCISES

1. Try smaller and larger values in the call **range(100)** on line 16 of Example 1.5. Describe the effects of each.

2. Try smaller and larger values in the call **move(10)** on line 17 of Example 1.5. Describe the effects of each.

3. Evaluate these boolean expressions:

   (a) `23 == 9 + 15`   (c) `30 >= 18 * 2`

   (b) `2 * 4 > 2 + 4`   (d) `5 * 4 <= 20`

4. Evaluate these boolean expressions:

   (a) `29 != 9 * 2`   (c) `15 - 4 >= 21`

   (b) `3 * 13 < 33`   (d) `38 <= 2 * 19`

5. Determine which block(s) will be executed in this code fragment:

```
Block A
if 20 < 10:
      Block B
Block C
```

6. Determine which block(s) will be executed in this code fragment:

```
Block A
if 5 >= 11 - 7:
    Block B
if 2 != 11 - 8:
    Block C
Block D
```

7. Look in the Python documentation to see if the turtle **shape()** function has a return value.

8. Look in the Python documentation to see if the turtle **speed()** function has a return value.

9. Find the one function used in Example 1.4 on page 18 that has a return value. Describe how the return value is used.

10. Describe what happens if **xcor()** is called in a program on a line by itself:

```
xcor()
```

If you are not sure, try adding it to a program.

11. Using **xcor()** and **ycor()**, write a **square(width)** function to draw a square so that the lower-left corner is put wherever the turtle is located, by only moving the turtle with **goto()**. Use your function to draw some squares.

12. Using **xcor()** and **ycor()**, write a **rectangle(width, height)** function to draw a rectangle so that the lower-left corner is put wherever the turtle is located, by only moving the turtle with **goto()**. Use your function to draw some rectangles.

13. Add another **if** statement to the **move()** function in Example 1.5 so that the turtle bounces off both the right and left sides.

14. Modify Example 1.5 so that the turtle bounces up and down, off the top and bottom, instead of moving side-to-side.

⟶ Note: It will be better to wait until Section 1.7 to bounce off all four sides. See the exercises there.

15. Modify Exercise 13 so that the turtle is blue when it is on the right half of the window and red when it is on the left half.

16. Write a program so that the turtle turns left when it gets close to any edge of the turtle window. Have it turn a different angle at each edge. Experiment with the angles to try to keep the turtle on screen.

1.6   CONDITIONAL REPETITION

A **for** loop repeats over a sequence, but there are situations where it is helpful to repeat based on a true-false condition rather than a specific number of times. The Python **while** loop combines the ability of an **if** statement to test a condition with the ability of a **for** loop to repeat. In the **spiral()** function of Example 1.6, the **while** loop keeps moving the turtle as long as the step size is positive.

```
1  # spiral.py
2  # Draw spiral shapes
3
4  from turtle import *
5
6  def spiral(firststep, angle, gap):
7      """Move turtle on a spiral path."""
8      step = firststep
9      while step > 0:
10         forward(step)
11         left(angle)
12         step -= gap
13
14 def main():
15     spiral(100, 71, 2)
16     exitonclick()
17
18 main()
```

Example 1.6   Spiral.

Syntax: **while** Loops

The syntax of a **while** loop is the same as an **if** statement, with **while** in the place of **if**:

```
while condition:
    body
```

As with an **if** statement, the condition of a **while** statement is a boolean expression. When the loop runs, *condition* is evaluated, and if it is **True**, then *body* is executed. What is different is that after the body executes, *condition* is evaluated again, and if it is still **True**, then the body executes again. The body is repeatedly executed until the condition is **False**.

`while` Loop Flow of Control

A **while** statement changes the flow of control to repeat like a **for** loop using a boolean condition like an **if** statement. Fig. 1.9 reveals the similarity to both. Notice also how the only way to continue to the following statement is for the boolean condition to be false.

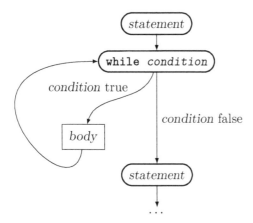

Figure 1.9 **while** loop flow of control.

Comparing `for` and `while` Loops

The major differences between **for** and **while** loops are:

- **for** loops repeat for every item in a sequence (with the value of that item available in a variable), whereas **while** loops repeat until a boolean condition fails, and

- **for** loops run a predictable number of times based on the number of items in their sequence, but usually the number of times a **while** loop will run is not known in advance.

These principles will help you choose between the two. In some cases, either loop can be used, and then the choice is a matter of preference. In fact, Example 1.6 could have been written with a **for** loop; see the exercises.

⟹ Caution: A **while** loop may be **infinite** if its condition never fails. Use Ctrl-C to stop the program or look for an option to restart your interpreter or shell if you execute an infinite loop.

An Introduction to Variables

Example 1.6 uses the variable `step` to control the turtle's step size as it walks the spiral. Think of variables in Python as names of values. Line 8 **assigns** the variable `step` to refer to the value in `firststep`:

```
step = firststep       # assign value of firststep to step
```

Line 12 updates the value `step` to decrease by the amount in `gap`:

```
step -= gap            # decrease step by value of gap
```

There is a corresponding `+=` operation to increase the value of a variable:

```
step += gap            # increase step by value of gap
```

Notice that when variables appear on the right-hand side of these statements, their values are accessed. Chapter 2 will explore these concepts in much more detail.

EXERCISES

1. List all of the values taken on by the `step` variable in Example 1.6 from the call `spiral(40, 30, 10)`. Of those values, indicate any that the turtle does not use to move forward.

2. List all of the values taken on by the `step` variable in Example 1.6 from the call `spiral(50, 28.1, 15)`. Of those values, indicate any that the turtle does not use to move forward.

3. List all of the values taken on by the `step` variable in each of the loops below. Indicate the value of `step` that stops the loop (if any).

   (a)
   ```
   step = 1
   while step < 10:
       step += 2
   ```
   (b)
   ```
   step = 50
   while step > 0:
       step -= 10
   ```

4. List all of the values taken on by the `step` variable in each of the loops below. Indicate the value of `step` that stops the loop (if any).

   (a)
   ```
   step = 3
   while step < 40:
       step += 5
   ```
   (b)
   ```
   step = 40
   while step > 0:
       step -= 7
   ```

5. Rewrite each loop in Exercise 3 as a **for** loop with a **range()**.

6. Rewrite each loop in Exercise 4 as a **for** loop with a **range()**.

7. Determine the number of times this loop body will execute:

   ```
   while 0 > 1:
       body
   ```

8. Determine the number of times this loop body will execute:

   ```
   while True:
       body
   ```

9. Rewrite Example 1.6 using a **for** loop instead of a **while** loop.

10. Experiment with the parameters of the `spiral()` function. Then write one program to draw three different spirals in different locations.

11. Rewrite the `main()` function of Example 1.4 to use a **while** loop instead of a **for** loop.

12. Rewrite the `polygon()` function of Example 1.4 to use a **while** loop instead of a **for** loop.

1.7   MORE COMPLEX CHOICES

Section 1.5 introduced **if** statements, which allow running code only when a boolean condition is true. Other forms of the **if** statement allow choosing from among two or more options.

Example 1.7 illustrates this with an adaptive circle function. When circles are drawn on pixel-based displays, they cannot be perfect, so they are usually drawn as polygons with many sides. The `mycircle()` function adapts the number of sides in the polygon to the radius, so that more sides are used for larger circles. It requires the `polygon()` function from Section 1.4.

```
1  def mycircle(radius):
2      """Draw circle as polygon."""
3      if radius < 20:
4          sides = 10
5      elif radius < 100:
6          sides = 30
7      else:
8          sides = 50
9      polygon(sides, 6.28*radius/sides)
```

Example 1.7   Adaptive circle.

Syntax: **else** and **elif** Clauses

A **clause** in Python is a piece of code that does not stand as a statement on its own but instead is appended to some other statement. An **if** statement may have an optional **else** clause containing alternate code to run when the condition is false:

```
if condition:
    body1
else:
    body2
```

In this case, *body1* is executed if the condition is true; otherwise, *body2* is executed. Think of this as an "either-or" choice, where one or the other of the two options will be chosen.

When the choice is more complex than either-or, a sequence of tests may be checked by using one or more **elif** clauses, again with an optional **else**:

```
if condition1:
    body1
elif condition2:
    body2
elif condition3:
    body3
...
else:
    bodyN
```

The keyword **elif** is short for "else if." If *condition1* is true, then *body1* is executed; otherwise, *condition2* is evaluated, and if it is true, *body2* is executed; and so on. Later tests are checked only if all preceding tests are false.

So, an **if-else** chooses between two options, whereas an **if-elif-else** chain chooses from any number of options.

Separate vs. Dependent Questions

Compare these two sequences of statements:

```
if score >= 90:          if score >= 90:
    # A                      # A
if score >= 80:          elif score >= 80:
    # B                      # B
```

The questions on the left are *separate*, and both questions will always be asked. For example, if **score** is 93, then both the *A* and *B* sections will run. The second question on the right is *dependent* on the answer to the first: **score >= 80** will only be tested if the first test fails. In this case, if **score** is 93, only the *A* section will run. Another way to think about the difference is to ask yourself if you need to run the second test when the first is true.

Use separate **if** statements to ask independent questions, each of which will always be asked; use an **if-elif** chain to choose one from among several options.

Boolean Operations

To express more complex conditions in any form of **if** or **while**, Python provides boolean **and**, **or**, and **not** operations. Table 1.7 describes how these operations work with boolean values.

TABLE 1.7   Boolean operations

x **and** *y*   True if both *x* and *y* are true; otherwise, **False**.
x **or** *y*   True if either one or both of *x* and *y* are true; otherwise, **False**.
not *x*   True if *x* is false; otherwise, **False**.

Example: xcor() >= 0 and xcor() < 300

> This expression has value **True** if the turtle's *x*-coordinate is both greater than or equal to 0 and less than 300. Both parts need to be true for an **and** to be true.

Example: Evaluate -25 >= 0 or -25 < 300.

> This expression has value **True** because at least one of the subexpressions, $-25 < 300$, is true.

⟹ Caution: Only use boolean values with these operations. They can be used with other types, but those idioms are more advanced, and the results may be unexpected.

EXERCISES

1. Draw a flow-of-control diagram like Fig. 1.7 for an **if-else**:

```
if condition:
    body1
else:
    body2
```

2. Draw a flow-of-control diagram like Fig. 1.7 for an **if-elif-else**:

```
if condition1:
    body1
elif condition2:
    body2
else:
    body3
```

3. Determine which block(s) will be executed in these code fragments:

(a) *Block A*
```
if 7 >= 3:
    Block B
else:
    Block C
Block D
```

(b) *Block A*
```
if 7 <= 29 - 14:
    Block B
if 14 != 2 * 6:
    Block C
Block D
```

4. Determine which block(s) will be executed in these code fragments:

(a) *Block A*
```
if 5 <= 11 - 7:
    Block B
elif 5 > 11 - 8:
    Block C
else:
    Block D
Block E
```

(b) *Block A*
```
if 7 <= 29 - 14:
    Block B
elif 14 != 2 * 6:
    Block C
Block D
```

5. Evaluate these boolean expressions:

(a) `10 >= 11 or 11 < 12`

(b) `3 < 1 and 42.5 >= 9 * 3`

6. Evaluate these boolean expressions:

(a) `(not 8 < 8) and 13 >= 13`

(b) `1.9 > 2 or 2 * 7 == 25 - 11`

7. Write the condition "xcor() is positive but less than 200" as a Python boolean expression.

8. Write the condition "ycor() is between -100 and -200 (inclusive)" as a Python boolean expression.

9. Incorporate Example 1.7 into a complete program that draws circles of radius 10, 20, 30, ..., 150.

10. Rewrite Exercise 14 of Section 1.5 using **else** and/or **elif** clauses, as appropriate.

11. Rewrite Exercise 15 of Section 1.5 using **else** and/or **elif** clauses, as appropriate.

12. Modify Example 1.5 from Section 1.5 so that the turtle bounces against all four sides. Start the turtle moving at an angle. Hint: bounce in the x direction by changing the heading to 180 - **heading()**; in the y direction, set it to -**heading()**.

13. Extend the previous exercise to draw a box in which the turtle will bounce. Do not worry too much about getting the turtle to bounce precisely against the box; just get it close.

14. Modify Example 1.5 from Section 1.5 so that the turtle wraps around when it reaches any edge of the window. For example, if the turtle is about to move off the right edge, it should reappear at the same point of the left edge. Start the turtle moving at an angle. Note: **speed(0)** is needed for this illusion to work.

1.8   RANDOMNESS

At this point, you have seen the core components of most software: functions, loops, conditions, and variables. As you continue to use these in different contexts, your facility will improve and your understanding deepen. The last two sections of this chapter provide more practice with turtles before moving on to numeric data in Chapter 2.

Random walks are not only fun to watch, they are important objects of study with applications across the sciences. Example 1.8 moves the turtle randomly on a grid, as long as it stays near the center of the screen. It uses the **randrange()** function from the **random** library to generate values that appear random. Table 1.8 lists it, along with some of the other functions available from this module.

Example 1.8 also uses **abs()**, Python's built-in absolute value function. It combines two comparisons in one, $|x| < 200 \iff -200 < x < 200$, to help test if the turtle is in the center of the screen.

Syntax: Importing Specific Names

To import only specific names from a module, list the names instead of using a star:

```
from module import name1, name2, ...
```

```
1  # randomwalk.py
2  # Draw path of a random walk.
3
4  from turtle import *
5  from random import randrange
6
7  def random_move(distance):
8      """Take random step on a grid."""
9      left(randrange(0, 360, 90))
10     forward(distance)
11
12 def main():
13     speed(0)
14     while abs(xcor()) < 200 and abs(ycor()) < 200:
15         random_move(10)
16     exitonclick()
17
18 main()
```

Example 1.8 Random walk.

For example, line 5 of Example 1.8 imports the **randrange** name from the **random** module. Use this form with most libraries, because otherwise, the *-form may accidentally import a name that conflicts with one of yours.

Nested Function Calls

Example 1.8 takes advantage of the fact that an argument to a function can come in many forms, as long as it represents an appropriate value. Consider the call on line 9 to **left(randrange(0, 360, 90))**. The argument to **left()** needs to be an angle, and in this case, the angle is given by the *return value* of

TABLE 1.8 **random** module: numeric functions

random()
Random value x with $0 \le x < 1$ (not necessarily an integer).
uniform(a, b)
Random value x with $a \le x \le b$ (not necessarily an integer).
randint(a, b)
Random integer n with $a \le n \le b$.
randrange(start, stop, step)
Random integer from **range(start, stop, step)**.

the call to `randrange()`. It may help to view this expression from the inside out, as in Fig. 1.10. Such an expression is known as a **nested function call**, because the inner call is nested inside the outer.

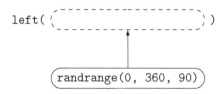

Figure 1.10 Nested function call.

EXERCISES

1. Describe the region of the window to which the turtle is restricted by the `while` loop on line 14 of Example 1.8.

2. Can Example 1.8 be rewritten with a `for` loop instead of a `while` loop? Explain why or why not.

3. List the names that would need to be listed if the *-form of `import` were not used on line 4 of Example 1.8.

4. List the possible values from which the `randrange()` expression on line 9 of Example 1.8 will choose.

5. Write a function call to return a random integer between 1 and 100.

6. Write a function call to return a random odd integer between 1 and 99.

7. Write a function call to return a random angle between 45 and 90 degrees (not necessarily an integer).

8. Write a function call to return a random angle between 0 and 360 degrees (not necessarily an integer).

9. Identify the two other nested function calls besides line 9 in Example 1.8, and draw a diagram similar to Fig. 1.10 for each.

10. Rewrite line 14 of Example 1.8 to make the same tests without using the `abs()` function.

11. Modify Example 1.8 so that the turtle makes turns that are multiples of 60 degrees.

12. Modify Example 1.8 so that the turtle makes turns of any angle between 0 and 360 (not restricted to integers).

13. Modify Example 1.8 so that the turtle takes a random number of steps between 1 and 20.

14. Modify Example 1.8 so that the turtle moves forward a random amount and makes random turns of any angle. Do not restrict either value to integers.

15. Write a program to draw 50 random squares in the turtle window, with random center and width. Use a function to draw each square.

16. Write a program to draw 50 random rectangles in the turtle window, with random center, width, and height. Use a function to draw each rectangle.

17. Write a program to draw 20 random circles in the turtle window, with random center and radius.

18. Write a `goto_random()` function that moves the turtle to a random location in the default turtle window. Use the function to draw something.

19. Write a program to draw 50 random line segments in the turtle window. Use a random thickness for each line.

1.9 THINKING WITH FUNCTIONS

Example 1.9 is the beginning of a program to draw faces using functions. It requires the `circle_at()` function from Section 1.3. This section will give you practice in using functions to help solve problems.

Tasks and Subtasks

Functions help solve complex problems by allowing us to think in terms of tasks and subtasks. The basic principle is that *each function should be designed to accomplish one well-defined task*. In doing so, it may call on other functions as subtasks in order to carry out its primary goal.

One way to help decide what various functions ought to do is by **separating responsibilities** between them. For instance, the functions in Example 1.9 isolate these responsibilities:

> **The face** is responsible for drawing its outer shape (a circle), and deciding where to put the eyes. If other components such as a nose or mouth are included, the face would be responsible for placing them.
>
> **The eye** is responsible for drawing its shape. If it had subcomponents, it would decide where to locate them.

So the `face()` function decides where to put the eyes, but it does not decide exactly what is drawn for each eye—that is the responsibility of the `eye()` function.

```
 1  # face.py
 2  # Draw a face using functions.
 3
 4  from turtle import *
 5
 6  # Include definition of circle_at()
 7
 8  def eye(x, y, radius):
 9      """Draw an eye centered at (x, y) of given radius."""
10      circle_at(x, y, radius)
11
12  def face(x, y, width):
13      """Draw face centered at (x, y) of given width."""
14      circle_at(x, y, width/2)
15      eye(x - width/6, y + width/5, width/12)
16      eye(x + width/6, y + width/5, width/12)
17
18  def main():
19      face(0, 0, 100)
20      face(-140, 160, 200)
21      exitonclick()
22
23  main()
```

Example 1.9 Draw a face.

Relative Locations and Sizes

Locations are determined relative to the specified center point of each component. For example, if the face's center is (x, y), then we can draw a picture like Fig. 1.11 to decide where to locate the eyes in terms of the point (x, y) and the entire *width* of the face.

Scope of Parameters

In programs like Example 1.9, it is important to realize that the x and y parameters in the face() function are completely different from the x and y parameters in the eye() function. They happen to have the same names, but that is all. The face parameters specify the center of the entire face, whereas the eye parameters are the center of an individual eye.

What makes this work in Python is a concept called scope. The **scope** of any variable is the region in a program where the variable may be used. A parameter's scope is the body of the function in which it is defined. That

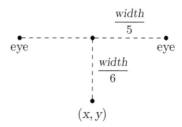

$$\frac{width}{5}$$

eye .... • .... eye

$$\frac{width}{6}$$

(x, y)

Figure 1.11 Determining relative locations.

means, for example, that the x parameter in `face()` cannot be used outside of the body of the `face()` function. Scope is convenient, because it means we do not need to create artificially different variable names.

Top-Down and Bottom-Up Design

Top-down design starts with large process steps and gradually breaks each one down until there is enough detail to implement it. Bottom-up works in reverse, beginning with relatively detailed tasks, and then putting those together until there is a complete solution.

In the context of Example 1.9, a top-down design would start with the face and decide what and where its subcomponents should be. A bottom-up approach, on the other hand, might design the eye, nose, and mouth before putting them together to make a face. Neither technique is always better, and in fact, it can be helpful to move back and forth between the two—particularly when you are stuck.

Defensive Programming

Look again at the `circle_at()` function from Example 1.3 on page 14. It makes sure the pen is up before moving to the starting point, it sets the pen down before drawing, and it makes sure the turtle is pointed in the right direction before calling `circle()`. In other words, `circle_at()` works no matter what direction the turtle is pointing or whether the pen is up or down before being called. This makes it easier to use, because you don't have to remember to set things in a certain way before calling it.

This is a type of **defensive programming** because the function defends itself against what a caller might or might not have done beforehand. It is a helpful technique when writing functions for subtasks.

EXERCISES

1. Write a helper `line(x0, y0, x1, y1)` function to draw the line segment between $(x0, y0)$ and $(x1, y1)$.

2. Write a `nose(x, y, radius)` function to draw a circular nose, and then add it to the face.

3. Write a `nose(x, y, height)` function to draw a straight-line nose, and then add it to the face.

4. Write a `mouth(x, y, width)` function to draw a straight-line mouth, and then add it to the face.

5. Modify the `eye()` function to add an inner filled eyeball.

6. Modify the `eye()` function to add an inner filled eyeball placed in a random location inside the eye.

7. Write a program to draw a line of faces horizontally across the screen.

8. Write a loop for y inside of a loop for x (called nested loops) to fill the screen with faces.

9. Write a program to draw faces in random locations and of random sizes.

10. Give your faces a body by developing a stick figure function. Include separate functions for the body, arms, and legs. Use your function to draw several stick figures of different sizes.

11. Write a function to draw a house based on the principles of this section. Include separate functions for components such as the base, roof, door, and windows. Use your function to draw several houses of different sizes.

12. Create your own drawing based on the principles of this section. Write separate functions for different subcomponents.

CHAPTER 2

Numeric Data

In this chapter, we shift from turtle graphics to handling numeric information in Python.

2.1 VARIABLES AND ASSIGNMENT

Example 2.1 shows one way to implement our own version of the built-in **max()** function[1] described in Table 2.1. It illustrates several important new concepts.

```python
# mymax2.py
# Write max function without built-in max().

def mymax2(x, y):
    """Return larger of x and y."""
    largest_so_far = x
    if y > largest_so_far:
        largest_so_far = y
    return largest_so_far

def main():
    print("MyMax: Enter two values to find the larger.")
    first = float(input("First value: "))
    second = float(input("Second value: "))
    print("The larger value is", mymax2(first, second))

main()
```

Example 2.1 Maximum.

---

[1] The built-in function works with more than two values, but we ignore that for now.

TABLE 2.1 Numeric functions

abs(x)
Return the absolute value of x, $|x|$.
max(x, y)
Return the larger of x and y.
min(x, y)
Return the smaller of x and y.
pow(x, y)
Return x raised to the power y, x^y.

Variables

As we saw in Chapter 1, variables allow us to keep track of and manipulate data in programs. In Python, a **variable** is a name that refers to a value during program execution. Different variable names allow a program to keep track of different values as the program runs. In this chapter, we will primarily use variables referring to numeric values.

Python variable names may be any legal identifier that is not a keyword (see page 15), but there are guidelines for choosing good variable names:

- Names should be meaningful in their context.

- Use lowercase.

- Combine words with underscores, as in `largest_so_far`.

Context matters with variable names, and different languages encourage different styles, as do different work environments. In all cases, good variable names help make programs easier to understand and maintain.

Syntax: Assignment

Variables are assigned values by **assignment** statements. The syntax of an assignment statement is:

```
variable = value
```

Think of assignment as attaching the variable's *name* to a *value*, like a nametag. The value must be determined before being named, so read assignments as two steps, from right to left:

1. Evaluate the expression on the right.

2. Assign the variable on the left to refer to that value.

⟹ Caution: An assignment "=" does not mean "equals." Use the comparison operator == to test whether two items are the same.

⟹ Caution: Variables must be assigned a value before use. Attempting to access the value of a variable that has not been assigned yet will generate a "name is not defined" NameError.

Visualizing Assignment

View assignments as creating an arrow from the variable to the value, and follow the arrow to determine the value of the variable when it is needed later. For example, consider this sequence of assignments:

```
x = 17
y = x + 10
```

The first assignment creates an arrow from x to the value 17, as in the top of Fig. 2.1. Then, in reading the right-hand side of the second statement, the arrow for x is followed to retrieve the value 17. Then 17 and 10 are added together, and y is assigned to the value 27.

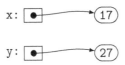

Figure 2.1 Assignment.

Local Variables in Functions

Variables used for the first time inside the body of a function are called **local variables** because they may only be used locally inside that function. In Example 2.1, the variable largest_so_far is local to the function mymax2(). The scope (see page 39) of a local variable is the body of the function in which it is defined—the same as the scope of function parameters. Any attempts to access a local variable from outside the function definition will fail; for an example, see Exercise 1. Thus, functions use a different mechanism—their return value—to communicate data back to the caller.

Syntax: **return** Statements

Recall from Section 1.5 that functions may have a return value. The mechanism to send a value from a function back to its caller is the **return** statement, which looks like this:

```
return value
```

When this statement is executed, it immediately terminates the function and returns the *value* as the value of the function call. Example 2.1 illustrates how to use a **return** statement in the function mymax2().

Because **return** statements immediately terminate the function, they often appear as the last statement in a function. However, **early returns** can also be useful in cases where the function can determine, sometime before its end, what the return value should be.

The *value* in a **return** is optional. Thus, **return** statements may be used in any function—for example, to terminate early. What is somewhat unusual in Python is that if no value is specified in a **return** statement, the special value None is returned. In fact, None is returned at the end of a function call even if you don't include a return statement! This means that *all* Python functions return something, even if that something is None.

Input and Output

Example 2.1 demonstrates basic text input and output using the built-in commands from Table 2.2. The **input()** function requests information from the user running the program, while **print()** outputs values to the interpreter window. Even though the input prompt is optional, use it so the user knows that the program expects a value. Values output with **print()** are separated by single spaces in the output, and each call of **print()** produces output on a separate line.

TABLE 2.2   Input and output

input(*prompt*)
Print optional *prompt* and return value input by user as a string.
print(*value1, value2, ...*)
Print *value1, value2*, etc., to screen, separated by spaces.

An Introduction to Types

Every data element in a Python program has a specific **type** that determines the operations the item supports as well as its range of possible values. Table 2.3 lists four of the most common types. In it, the term "float" is short for "floating point," which refers to numbers that use a decimal point. Start becoming aware of these types, because if you try to perform an operation on a value of the wrong type, it will either fail or produce unexpected results. In particular, numeric computations cannot be done with strings, and there will be important differences between integers and floats that we encounter throughout this chapter.

Strings written with double quotes may contain single quotes, and vice versa. Triple-quoted strings may extend across multiple lines.

TABLE 2.3 Basic types

`bool`	True or `False`
`float`	Numeric values with a decimal point such as 3.141, -23.8, 7.0, or `5.92e7` $= 5.92 \times 10^7$
`int`	Whole number value such as 847, -19, or 7
`str`	String of characters inside single, double, or triple quotes, such as `'abc'`, `"It's here!"`, or docstring comments (see page 15)

Converting Types

The `input()` function always returns user input as a string, so in order to process numeric input, the string must be converted to either an integer or float. Table 2.4 lists some of the type-conversion functions in Python; each one is simply the name of the type followed by parentheses. For the input of numeric data, choose between `int()` and `float()` depending on the context.

TABLE 2.4 Type converters

`float(`x`)`
Convert x to a floating-point value if possible.
`int(`x`)`
Convert x to an integer if possible, discarding any fractional part.
`str(`x`)`
Convert x to a string.

$\Longrightarrow$ Caution: `int(1.72)` returns 1, but `int("1.72")` generates an error because the string does not contain a valid integer.

EXERCISES

1. Change Example 2.1 to print `largest_so_far` in `main()`, and run the program. Describe what happens.

2. Change Example 2.1 to use `int()` instead of `float()`, and run the program. Describe the differences.

3. Change Example 2.1 to delete the **return** statement (line 9) from the `mymax2()` function, and run the program. Describe and explain the result.

4. Variables can help unravel nested function calls. To see how, rewrite Example 2.1 to assign a name to the return value of the `input()` statement on line 13, and then convert that value to a float in a second step.

5. Determine the final values of each variable in these sequences of assignment statements, unless there would be an error during execution. In that case, explain why there is an error.

(a)
```
a = 23
b = a - 10
c = a + b
```

(b)
```
i = 4
j = i + k
k = i + 3
```

6. Draw pictures to help determine the final values of each variable in these sequences of assignment statements, unless there would be an error during execution. In that case, explain why there is an error.

(a)
```
p = 26
q = p - 11
r = p + q + r
s = r - p - q
```

(b)
```
w = 30
x = w + w
y = w - x
z = w + x - y
```

7. Determine the output of this code fragment:

```
x = 10
y = 15
print(x, y)
y = x
x = y
print(x, y)
x = 5
print(x, y)
```

8. Determine the output of this code fragment:

```
x = 10
y = 15
z = 20
x = z
z = y
y = x
print(x, y, z)
```

9. Rewrite the `mymax2()` function of Example 2.1 without using a local variable.

10. Suppose x and y have been assigned values. Write assignment statements to **swap** their values, so that after your statements run, x has the original value of y, and y has the original value of x.

11. Write a `mymin2(x, y)` function to return the smaller of two values without using the built-in `min()` function. Include a `main()` to test your function.

12. Write a `myabs(x)` function to return the absolute value of x, without using the built-in `abs()` function. Include a `main()` to test your function.

13. Rewrite the `main()` of Example 2.1 to use a loop for y inside of a loop for x to test `mymax2()` on all combinations of x and y with $0 \leq x, y \leq 4$.

14. Write a `mymin3(x, y, z)` function to return the smallest of three values without using the built-in `min()` function. Include a `main()` to test your function.

15. Write a `mymax3(x, y, z)` function to return the largest of three values without using the built-in `max()` function. Include a `main()` to test your function.

16. Write a `median3(x, y, z)` function to return the middle value among x, y, and z when listed in increasing order. If two or more of the values happen to be the same, that value is the median. Include a `main()` to test your function.

17. Write a function `grade(score)` to return the corresponding letter grade for a given numerical score. That is, for 90 or above, return the string `"A"`, for 80 or above, return `"B"`, etc. Include a `main()` to test your function.

2.2 CALCULATIONS

High-level programming languages like Python are a powerful tool for complex or repetitive calculations. Example 2.2 shows how to create a table of values for a bank account earning compound interest. It uses the basic arithmetic operations in Table 2.5 and a `for` loop to make the table.

TABLE 2.5 Arithmetic operations

`x + y`   Sum of x and y.
`x - y`   Difference of x and y.
`x * y`   Product of x and y.
`x / y`   Quotient of x and y. Result is a float.
`x // y`   Integer quotient of x and y, rounding down.
`x % y`   Remainder after integer division of x by y, also known as mod.
`x ** y`   x to the power y, x^y. Same as `pow(x, y)`.

```
1  # balance_table.py
2  # Print table of account balances earning interest.
3
4  def balance(p, r, t):
5      """Return new balance using compound annual interest."""
6      return p*(1 + r)**t
7
8  def main():
9      print("Calculates compound interest over time.")
10     principal = float(input("Principal: "))
11     rate = float(input("Interest rate (as a decimal): "))
12     years = int(input("Number of years: "))
13     for year in range(years + 1):
14         print(year, balance(principal, rate, year))
15
16 main()
```

Example 2.2   Table of account balances.

Order of Operations

Calculations in Python follow the normal rules of arithmetic. These include rules for deciding the order of operations, based on **operator precedence**. This means that the arithmetic operations are arranged in a hierarchy, part of which is shown in Table 2.6. Operators of higher precedence are computed first, while operators of equal precedence are performed left-to-right. Parentheses may always be used to group terms.

TABLE 2.6   Operator precedence

Higher	Power	**
	Multiplicative	*, /, //, %
Lower	Additive	+, −

Example: Evaluate 1 + 2 * 3 - 8 / 2.

> The solution is 3. In this expression, the multiplication 2 * 3 and division 8 / 2 are computed first, and then those results are combined left-to-right, as $(1 + 6) - 4 = 7 - 4 = 3$.

Leaving out the spaces around multiplicative and power operators can help remind readers that these bind more tightly, as in 1 + 2*3 - 8/2. This style is used in line 6 of Example 2.2.

Integer Operations

Keep in mind that **int** and **float** are different numeric types in Python. Two of the operations in Table 2.5 are specifically designed for integers: the integer quotient (//) and remainder (%). Integer quotient is the usual quotient rounded down to the nearest integer, while the remainder is the integer remaining after computing the integer quotient.

Example: 17 // 5 is 3.

The quotient $17/5 = 3.4$ rounds down to the integer 3. Another way to think of integer quotient for positive values is the number of times the integer 5 can fit completely inside of 17, which is 3.

Example: 17 % 5 is 2.

The remainder left after dividing 17 by 5 is 2. For positive values, think of this as the number left over after taking out as many whole 5's as possible.

Example: For integer **n**, **n** % 2 indicates whether **n** is even or odd.

If **n** % 2 is 0, then **n** is even; otherwise, if **n** % 2 is 1, **n** is odd.

The math Module

Python offers many other mathematical operations and constants in the **math** library module. Some common ones are given in Table 2.7; check the documentation for a complete list.

TABLE 2.7   **math** module: basic functions

sqrt(x)
Square root of x, $\sqrt{x}$.
floor(x)
Largest integer less than or equal to x, the floor of x.
ceil(x)
Smallest integer greater than or equal to x, the ceiling of x.
log(x)
Natural logarithm of x, $\ln x$.
exp(x)
Exponential e^x.
pi
The constant π.

Example: `floor(-3.14)` is −4, and `ceil(3.14)` is 4.

Compare how these work to the **int()** type converter: **int()** always discards the fractional part.

Example: `m//n` is identical to `floor(m/n)`.

This provides another way to think about how `//` works. It also explains why `//` is also called the floored quotient.

EXERCISES

1. Determine the value of each of these expressions:

   (a) `1 + 2 * 3 - 4 * 5 + 6`   (f) `12 <= 3**3`

   (b) `1 + 2 ** 3 * 4 - 5`   (g) `7 % 3`

   (c) `7 // 3`   (h) `244 % 2`

   (d) `15 // 4`   (i) `floor(15.4)`

   (e) `1.4**2 > 2`   (j) `ceil(8.229)`

2. Determine the value of each of these expressions:

   (a) `1 * 2 - 3 * 4 + 5 - 6`   (f) `5/8 != 5//8`

   (b) `1 / 2 - 3 / 4`   (g) `6 % 8`

   (c) `6 // 8`   (h) `1011 % 2`

   (d) `10 // 12`   (i) `ceil(23.18)`

   (e) `6//7 == 1//7`   (j) `floor(-20.48)`

3. Determine the value of each of these expressions:

   (a) `(1 + 2) * (3 - 4) * 5 + 6`   (f) `7/2 == 3 or 1 >= 8/5`

   (b) `1 * 4 - 2 / 2`   (g) `11 % 2`

   (c) `11 // 2`   (h) `988 % 5`

   (d) `5 / 2 - 5 // 2`   (i) `floor(-7.03)`

   (e) `10 >= 11 or 11 < 12`   (j) `ceil(-17.05)`

4. Determine the value of each of these expressions:

   (a) `1 - 2 * 3 + (4 + 5 * 6)`   (f) `1 == 5//3 and not 18 < 11`

   (b) `(1 + 4 - 2) / 2`   (g) `9 % 7`

   (c) `9 // 7`   (h) `1543 % 10`

   (d) `23 // 6 + 14 // 3`   (i) `ceil(-4.999)`

   (e) `3 > 1 and 4.25 > 9/2`   (j) `floor(5.661)`

5. List the operations in the order they are performed in the expression `p*(1 + r)**t`.

6. List the operations in the order they are performed in the expression `p*1+r**t`.

7. Write a Python expression to compute $m = \dfrac{y2 - y1}{x2 - x1}$.

8. Write a Python expression to compute $y = ax^2 + bx + c$.

9. Modify Example 2.2 to compound monthly. Use the formula $P(1 + \dfrac{r}{12})^{12t}$.

10. Modify Example 2.2 to ask the user for the number of compounding periods per year, n. Compute the new balance with the formula $P(1 + \dfrac{r}{n})^{nt}$.

11. Modify the `main()` of Example 2.2 to compute for a fixed number of years but loop over different principal values. Ask the user for the interest rate and number of years, and then compute the compound interest for principals ranging from $1000 to $10,000.

12. Modify the `main()` of Example 2.2 to compute for a fixed number of years but loop over different interest rates. Ask the user for the principal and number of years, and then compute the compound interest for interest rates from 1% to 10%. Hint: `range()` only produces ints, so start with a range of integers 1 through 10.

13. Modify the `main()` of Example 2.2 to display account growth over time for different interest rates. Ask the user for the principal, and then for each interest rate from 1% to 10%, display the balance after years 1 through 10.

14. Change the perspective in the previous exercise by displaying, for each time period from 1 year to 10 years, the balance earned by each interest rate from 1% to 10%.

15. Write a program to print a table of values of n and 2^n for $n = 1, 2, \ldots, 10$. You do not need any functions other than `main()`.

16. Write a program to print a table of values of n, $\log n$, n^2, and 2^n for $n = 10, 20, \ldots, 200$. You do not need any functions other than `main()`.

17. Write a program to print a table of the areas of circles of radius $r = 1, 2, \ldots, 10$. Write a function to compute the area, which is πr^2.

18. Write a program to print a table of the volumes of spheres of radius $r = 1, 2, \ldots, 10$. Write a function to compute the volume, which is $\dfrac{4}{3}\pi r^3$.

19. Write a program to print a table of Fahrenheit to Celsius conversions for temperatures between $-30°F$ and $100°F$ at 10-degree intervals. Write a function to convert Fahrenheit to Celsius; the formula is $\frac{5}{9}(F - 32)$.

20. Write a program to print a table of mile-to-kilometer conversions for distances between 100 and 1500 miles at 100-mile intervals. Write a function to do the conversion; one mile is approximately 1.609 km.

21. The formula $208 - 0.7y$ is an estimate of a person's maximum heart rate in beats per minute when they are y years old. Write a program to print a table of values of these estimates for ages between 20 and 60 at 2-year intervals. Write a function to compute the estimate.

22. (Requires Exercise 21.) Write a function to return a description of a person's training zone based on his or her age and training heart rate, rate. The zone is determined by comparing rate with the person's maximum heart rate m: a rate at least 90% of m is considered "interval training"; otherwise, 70% of m is "threshold training"; and 50% of m is "aerobic training". Any rate below 50% is classified as "couch potato". Use the function from the previous exercise to estimate m, and write a main() to ask the user for input and display the result.

23. One way to approximate $\sqrt{x}$ is to start with an initial guess g and then repeatedly replace g with the average of g and x/g. Use this method to approximate $\sqrt{3}$ by hand.

24. Implement the idea from the previous exercise as a function mysqrt(x). Use $g = x/2$ as the initial guess, and repeat until g^2 is within 10^{-10} (written 1e-10 in Python) of x. Compare your function with the sqrt() function.

2.3   ACCUMULATION LOOPS

Many types of computation involve repeated accumulation. In fact, compound interest is one example: interest gradually accumulates over time in a bank account. However, there was no sign of accumulation in the balance() function of Example 2.2 because it used a direct formula, $P(1 + r)^t$, to compute each balance.

Example 2.3 shows how to rewrite the balance() function so that it repeatedly accumulates compound interest over time instead of using the direct formula. This function can take the place of balance() in Example 2.2 to produce the same table of values: just change main() to call balance_accum() instead of balance().

```
1  def balance_accum(principal, rate, years):
2      """Return balance with accumulated compound interest."""
3      balance = principal
4      for _ in range(years):
5          interest = balance * rate
6          balance += interest
7      return balance
```

Example 2.3 Accumulating interest.

Assignment Shorthands

Example 2.3 uses the operation +=, which we saw briefly in Chapter 1. Think of it as *adding to* a variable. For example, this statement on line 6:

    balance += interest

means to add interest to the existing balance. In other words, it is a shorthand for this assignment:

    balance = balance + interest

Table 2.8 lists some of the shorthands that are useful for accumulation. Use them whenever you can: they are usually more efficient that the full assignment statements.

TABLE 2.8   Assignment shorthands

x += y
Add y to x. Similar to x = x + y.
x -= y
Subtract y from x. Similar to x = x - y.
x *= y
Multiply x by y. Similar to x = x * y.
x /= y
Divide x by y. Similar to x = x / y.

See the Python documentation for a complete list, including //= and %=, where they are also called **augmented** assignments.

Accumulation Loops

The **for** loop in Example 2.3 is called an **accumulation loop**. In an accumulation loop, one of the variables, known as the **accumulator**, gradually builds up some value as the loop runs. In this case, the accumulator is the variable **balance**. It begins with the value of the **principal** in line 3, and then each time the loop runs, the step in line 6 adds another year's interest onto the balance.

Accumulation loops follow this pattern:

```
accumulator = starting value
loop:
    accumulator += amount to add
```

After the loop finishes, the *accumulator* variable contains the accumulated value. As in Example 2.3, the loop may contain other code in addition to the += accumulation step.

Example:

```
result = 0                              0 1
for i in range(4):        output        1 2
    result += 1            ──────→       2 3
    print(i, result)                     3 4
```

The variable i gets its value from the **range()**, while result increases by 1 each time the loop runs.

⟹ Caution: Do not use **sum** as an accumulator variable name, because it is a built-in Python function.

Accumulation loops may also accumulate via multiplication rather than addition. In this case, the pattern is:

```
accumulator = starting value
loop:
    accumulator *= amount to multiply on
```

Assignment Precedence

Assignment and assignment shorthands have low precedence in Python, which means you can do calculations on the right-hand side of a shorthand and be sure that they will be computed prior to the assignment. In other words,

```
x += 2 * 3 - 1
```

is equivalent to:

```
x += 5
```

because 2 * 3 - 1 is evaluated before the assignment.

EXERCISES

1. Determine the final value of the `result` accumulator after each of these accumulation loops:

   (a)
   ```
   result = 0
   for _ in range(5):
       result += 1
   ```

   (b)
   ```
   result = 0
   for i in range(5):
       result += i
   ```

2. Determine the final value of the `result` accumulator after each of these accumulation loops:

   (a)
   ```
   result = 0
   for _ in range(5):
       result += 2
   ```

   (b)
   ```
   result = 0
   for i in range(5):
       result += 2*i
   ```

3. Show the output of each of these code fragments:

   (a)
   ```
   result = 0
   for i in range(5):
       print(i, result)
       result += 2
   ```

   (b)
   ```
   result = 1
   for j in range(5):
       result *= 2
       print(j, result)
   ```

4. Show the output of each of these code fragments:

   (a)
   ```
   result = 0
   for j in range(1, 10, 2):
       result += j
       print(j, result)
   ```

   (b)
   ```
   result = 0
   for k in range(5):
       print(k, result)
       result += 2 * k + 1
   ```

5. Write an accumulation loop to compute each of these quantities:

   (a) $1 + 2 + 3 + \cdots + 100$

   (b) $5 + 10 + 15 + \cdots + 100$

   (c) $1 + \dfrac{1}{2} + \dfrac{1}{3} + \cdots + \dfrac{1}{100}$

6. Write an accumulation loop to compute each of these quantities:

   (a) $2 + 4 + 6 + \cdots + 100$

   (b) $1 + 4 + 9 + 16 + \cdots + 100$

   (c) $1 + \dfrac{1}{4} + \dfrac{1}{9} + \cdots + \dfrac{1}{100}$

7. Write an accumulation loop to compute $20! = 1 \cdot 2 \cdot 3 \cdots \cdot 20$.

8. Write an accumulation loop to compute $2^{20} = 2 \cdot 2 \cdot 2 \cdots \cdot 2$.

9. Modify Example 2.3 to compound monthly. Interest is earned every month, but at $1/12$ the annual rate.

10. Modify Example 2.3 to compound interest n times per year, using $1/n$ times the annual rate. Ask the user for the number of compounding periods, n.

11. Write a function `triangular(n)` to return the sum $1 + 2 + 3 + \cdots + n$. (These sums are known as triangular numbers.) Write a program to print a table of values of the triangular numbers for $n = 1, 2, 3, \ldots, 20$.

12. Write a function `harmonic(n)` to return the sum $1 + \dfrac{1}{2} + \dfrac{1}{3} + \cdots + \dfrac{1}{n}$. (These sums are known as harmonic numbers.) Write a program to print a table of values of the harmonic numbers for $n = 1, 2, 3, \ldots, 20$.

13. Write a function `sum_of_evens(n)` to return the sum of the first n even integers, $2 + 4 + 6 + \cdots + 2n$. Write a program to print a table of values of this function for $n = 1, 2, 3, \ldots, 20$.

14. Write a function `sum_of_odds(n)` to return the sum of the first n odd integers, $1 + 3 + 5 + \cdots + (2n - 1)$. Write a program to print a table of values of this function for $n = 1, 2, 3, \ldots, 20$.

15. Write a function `factorial(n)` to return $n! = 1 \cdot 2 \cdot 3 \cdots \cdots n$. Write a program to print a table of values of `factorial()` for $n = 1, 2, 3, \ldots, 20$.

16. Write a function `pyramid(n)` to return the sum $1^2 + 2^2 + 3^2 + \cdots + n^2$. (These sums are known as square pyramidal numbers.) Write a program to print a table of values of the pyramidal numbers for $n = 1, 2, 3, \ldots, 20$.

17. Write a function `baselsum(n)` to return the sum $1 + \dfrac{1}{2^2} + \dfrac{1}{3^2} + \cdots + \dfrac{1}{n^2}$. Write a program to print a table of values of these sums for $n = 1, 2, 3, \ldots, 20$.

2.4   ACCUMULATOR OPTIONS

Section 2.3 describes a general pattern of accumulation that will continue to recur in different contexts. This section explores several variations on the theme of accumulation that are especially useful with numeric data. Example 2.4 illustrates three of them by asking a natural question: how long does it take an account earning interest to reach a savings goal? While there are mathematical ways to answer this question, a program can simply accumulate interest until the goal is reached.

```
1  def years_to_goal(principal, rate, goal):
2      """Return number of years to reach savings goal."""
3      balance = principal
4      years = 0
5      while balance < goal:
6          interest = balance * rate
7          balance += interest
8          years += 1
9      return years
```

Example 2.4 Savings goal.

Accumulating with a `while` Loop

Example 2.3 in the previous section is written with a **for** loop, but that will not help us solve this new problem. A **for** loop fits a setting in which we are given the number of years to accumulate interest, but here that is exactly what we do not know. A **while** loop is more appropriate for this new problem, repeating until the balance reaches the goal.

The accumulation pattern works with any loop, **while** or **for**, allowing you to choose based on the problem and context. That is why it is written with a generic "loop:"

```
accumulator = starting value
loop:
    accumulator += amount to add
```

Example 2.4 uses a **while** loop on line 5 to repeat as long as the balance is less than the goal. This is precisely what we need in order to determine the number of years it will take.

Counter Variables

The `years_to_goal()` function needs to count how many times its loop runs. We can do that with a particular type of accumulator, a **counter** variable. The simplest counter is just an integer variable that starts at 0 and increases by 1 every time the loop runs:

```
counter = 0
loop:
    counter += 1
```

Example 2.4 uses **years** as a counter variable.

If we allow for counting by different amounts, then a counter variable can produce any sequence that a **range()** produces.

Example: Create a counter variable to produce 2, 6, 10, ..., 50.

> This loop shows one way, although other boolean conditions could be used in the loop:
>
> ```
> n = 2
> while n <= 50:
>     n += 4
> ```

Usually, **for** loops do not need counter variables because they already have a variable driving the loop; counters are much more common with **while** loops.

Using Multiple Accumulators

To solve the problem of determining how many years it will take to reach a savings goal, both a balance accumulator and a counter variable are needed. It is important to recognize that the *same loop* can accumulate both: we do not need (or want) two separate loops.

Recognizing accumulators can also help you understand unfamiliar code. If we isolate the two accumulators in Example 2.4, then it is easier to figure out what the whole function is doing:

```
balance = principal             years = 0
while balance < goal:           while balance < goal:
    balance += interest             years += 1
```

The left set of statements accumulates the balance until it reaches the goal, while the right set counts the years—at the same time—because they use the same loop. Looking for familiar patterns like this will help develop your ability to read code.

Decreasing Loops

Although "accumulation" implies growth, similar loops may be written for quantities that count down or decrease. For example, payments may be repeatedly withdrawn from an account with a loop like this:

```
balance = principal
loop:
    balance -= payment
```

Here, `-=` is used instead of `+=` so that the balance decreases by the payment amount.

EXERCISES

1. Determine the final value of the **result** accumulator after each of these accumulation loops:

    (a) ```
    result = 0
    while result < 10:
        result += 3
    ```

    (b) ```
    result = 2
    while result < 10:
        result += result
    ```

2. Determine the final value of the **result** accumulator after each of these accumulation loops:

    (a) ```
    result = 10
    while result > 0:
        result -= 3
    ```

    (b) ```
    result = 0
    while result < 10:
        result += result + 1
    ```

3. Show the output of each of these code fragments:

    (a) ```
    i = 0
    result = 0
    while result < 10:
        i += 1
        result += i
        print(i, result)
    ```

    (b) ```
    j = 1
    result = 0
    while result < 10:
        result += j
        j *= 2
        print(j, result)
    ```

4. Show the output of each of these code fragments:

    (a) ```
    n = 1
    result = 0
    while result < 10:
        result += n
        n += 2
        print(n, result)
    ```

    (b) ```
    k = 1
    result = 0
    while k < 10:
        result += k
        k *= 3
        print(k, result)
    ```

5. Write an accumulation loop using **while** to compute each of these quantities:

    (a) $1 + 2 + 3 + \cdots + 100$

    (b) $5 + 10 + 15 + \cdots + 100$

    (c) $1 + \dfrac{1}{2} + \dfrac{1}{3} + \cdots + \dfrac{1}{100}$

6. Write an accumulation loop using **while** to compute each of these quantities:

    (a) $2 + 4 + 6 + \cdots + 100$

    (b) $1 + 4 + 9 + 16 + \cdots + 100$

    (c) $1 + \dfrac{1}{4} + \dfrac{1}{9} + \cdots + \dfrac{1}{100}$

7. Write an accumulation loop using **while** to compute $20! = 1 \cdot 2 \cdot 3 \cdots \cdots 20$.

8. Write an accumulation loop using **while** to compute $2^{20} = 2 \cdot 2 \cdot 2 \cdots \cdots 2$.

9. Give another test that could be used in the **while** loop example on page 60 to produce the sequence 2, 6, 10, ..., 50.

10. What does the **years_to_goal()** function in Example 2.4 return if it is called with a goal less than or equal to the principal? Explain.

11. Write a complete program using **years_to_goal()** from Example 2.4 to ask the user for a starting principal, interest rate, and savings goal. Then compute and print the number of years until the goal is reached.

12. Modify the **years_to_goal()** function to add a parameter for an annual deposit. Add the deposit each year after computing interest. Write a complete program to ask the user for the principal, interest rate, annual deposit, and goal, and then print the number of years until the goal is reached.

13. A loan is taken out with interest compounded annually, where equal annual payments will be made.

    (a) Write a **years_to_payoff(owed, rate, payment)** function to return the number of years that it will take to pay off the loan. Use it to write a program that asks the user for the amount owed, interest rate, and payment amount, and then prints the number of years it will take to pay.

    (b) If the annual payment is not greater than the amount of interest for the first year, the loan will never be paid. Add a test to report this to the user instead of calling the function.

14. A loan is taken out with interest compounded monthly, where equal monthly payments will be made.

    (a) Write a **years_to_payoff(owed, rate, payment)** function to return the number of years that it will take to pay off the loan. Use it to write a program that asks the user for the amount owed, annual interest rate, and monthly payment amount, and then prints the number of years it will take to pay.

    (b) If the monthly payment is not greater than the amount of interest for the first month, the loan will never be paid. Add a test to report this to the user instead of calling the function.

15. Money is deposited in an account with interest compounded annually from which equal annual withdrawals will be made until the account has a zero balance. Do not worry about the exact amount of the final withdrawal.

   (a) Write a function to return the number of years that withdrawals can be made given the starting balance, interest rate, and withdrawal amount. Use it to write a program that asks the user for these values and then prints the number of years the account will have money in it.

   (b) If the annual withdrawal is not greater than the amount of interest for the first year, the balance will never decrease. Add a test to report this to the user instead of calling the function.

16. Money is deposited in an account with interest compounded monthly from which equal monthly withdrawals will be made until the account has a zero balance. Do not worry about the exact amount of the final withdrawal.

   (a) Write a function to return the number of years that withdrawals can be made given the starting balance, interest rate, and withdrawal amount. Use it to write a program that asks the user for these values and then prints the number of years the account will have money in it.

   (b) If the monthly withdrawal is not greater than the amount of interest for the first month, the balance will never decrease. Add a test to report this to the user instead of calling the function.

17. Write the function `smallest_divisor(n)` to return the smallest integer $k > 1$ that divides evenly into n. For example, `smallest_divisor(25)` is 5. Testing whether or not $n \bmod k$ equals 0 will determine if k evenly divides n. Write a program to print a table of the smallest divisors of $n = 2, 3, 4, \ldots, 50$.

18. Write the function `intlog(n)` to return the largest power of 2 that is less than or equal to n; that, is the largest k with $2^k \leq n$. For example, `intlog(20)` is 4, and `intlog(32)` is 5. Use a counter rather than any library functions. Write a program to print a table of values of `intlog()` for $n = 1, 2, \ldots, 50$.

PROJECT: SIMULATION

One of computing's great strengths is its ability to run simulations of system models, often of great complexity, across a wide range of problem domains. A simple model of a fair coin is a random integer that is equally likely to be 0 or 1. This model can be used to answer questions such as, if a coin comes up heads 15 out of 20 times, how likely is it that the coin is fair?

The idea is to perform many trials of an experiment: flip a coin in the model 20 times, and count the number of heads. If it is 15 or more, register this trial as a success. In the end, the number of successes divided by the number of trials is an estimate of the likelihood that the coin is fair.

Nested Loops

Simulating multiple trials of a given number of flips will require **nesting** the loop for flips inside a loop over trials, as pictured in Fig. 2.2. Nesting causes the inner loop to run fully every time the outer loop runs, as needed here. The basic principle is that everything inside the outer loop is repeated, including any loops that happen to be there.

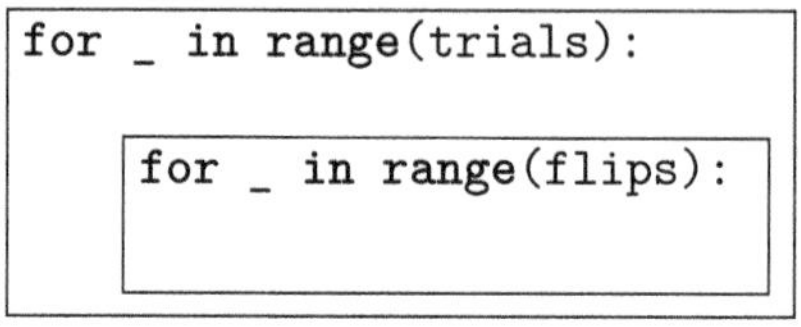

```
for _ in range(trials):

    for _ in range(flips):
```

Figure 2.2 Nested loops.

EXERCISES

1. Write a function `flip()` to simulate flipping a fair coin by randomly returning either 0 or 1. Include a program to test your function.

2. Use the `flip()` function of the previous exercise to write a program to perform the test of a coin described above. Ask the user for the number of times to flip, the minimum number of heads to count as success, and the number of trials. Report the number of trials with at least that many heads and the percentage of trials that were successful.

3. Write a function `roll()` to simulate rolling a fair six-sided die by randomly returning an integer between 1 and 6. Include a program to test your function.

4. Use the `roll()` function of the previous exercise to write a program to test the fairness of a die that rolls a six on 12 out of 30 rolls. Ask the user for the number of times to roll, the minimum number of sixes to count as success, and the number of trials. Report the number of trials with at least that many sixes and the percentage of trials that were successful.

5. A game show contestant stands in front of three large doors. Behind one of the doors is a new car, while the other doors conceal goats. The contestant chooses a door. The host then opens one of the other two doors to reveal a goat, and offers the player the option to switch to the remaining closed door or stay with his or her original choice. Write a simulation to help determine the best strategy for the player.

6. Write a program to simulate a coin-flipping game with two players, Same and Different. In each round, each player flips one coin, and if they are the same (both heads or both tails), then player Same wins both coins; otherwise, player Different takes both coins. Play continues until one player is out of coins. Ask the user for the starting number of coins for each player, and for each round, report the flips, who wins the round, and each player's new number of coins. Report the winner at the end.

7. Given what you already know about computer programs, explain why it is difficult for a program to compute "random" numbers. Research the meaning of the term **pseudorandom** and report what you find.

2.5 NUMBERS IN MEMORY

Now that you have seen some accumulation loops, predict the output of Example 2.5, and then run it. Are you surprised? Use Ctrl-C if you need to.

```
1  # loop.py
2  # A subtle accumulation loop.
3
4  def main():
5      x = 0
6      while x != 1:
7          print(x)
8          x += 0.1
9
10  main()
```

Example 2.5   A surprising loop.

To understand this loop and better understand Python arithmetic, we look at how numeric values are stored in memory.

Bits and Bytes

All computer memory is made up of bits. Without getting into electronics or physics, one **bit** can be thought of as a single on-off switch. A group of 8 bits is a **byte**. In abbreviations, a small "b" refers to bits, while capital "B" stands for bytes. For example, Mbps = megabits per second, and GB = gigabytes.

Binary Numbers

The next step toward understanding how numbers are stored is to interpret each bit using *Off* = 0 and *On* = 1. For example, one byte might be thought of this way:

Off	*On*	*On*	*Off*	*On*	*Off*	*Off*	*Off*
0	1	1	0	1	0	0	0

The key to working with numbers made up of only 0's and 1's is to interpret them in base two, also known as binary. To understand binary, it is helpful to go back to the meaning of decimal numbers.

> **Base 10 (decimal)** numbers use place values based on powers of 10. For example,

$$\boxed{2} \quad \boxed{1} \quad \boxed{7} \quad \boxed{4} \quad = 2*1000 + 1*100 + 7*10 + 4*1$$

$$\begin{array}{cccc} 1000\text{'s} & 100\text{'s} & 10\text{'s} & 1\text{'s} \\ 10^3 & 10^2 & 10^1 & 10^0 \end{array}$$

> The 2, for example, has a value of 2000 because it represents $2*10^3$.

> **Base 2 (binary)** numbers work the same way except they are based on powers of two instead of powers of ten, and they only use digits 0 and 1. For example,

$$\boxed{1} \quad \boxed{0} \quad \boxed{1} \quad \boxed{1} \quad = 1*8 + 0*4 + 1*2 + 1*1 = 11$$

$$\begin{array}{cccc} 8\text{'s} & 4\text{'s} & 2\text{'s} & 1\text{'s} \\ 2^3 & 2^2 & 2^1 & 2^0 \end{array}$$

> Here, the first 1 has a value of 8 because it represents $1*2^3$.

Integers in Memory

The basic idea, then, is that because all computer memory is composed of bits, integers are stored in binary. Describing a computer as having a "32-bit" or "64-bit" architecture tells you the basic memory size used by its CPU, which is normally the number of bits used to store integers.

There is one complication, though: we have not described how to store negative integers. Simple binary is only used for **unsigned** integers, which are greater than or equal to zero. **Signed** integers have a positive or negative sign, and they require a more complicated representation that is usually described in later courses.

Floats in Memory

Floating-point values are even more complicated to store in memory, but we can see enough here to understand what goes wrong with the loop in

Example 2.5. The basic idea is that fractional values are also stored in binary using negative powers of 2. For example,

$$0 \, . \, \boxed{1} \quad \boxed{0} \quad \boxed{1} \quad \boxed{1} \quad = 1 * \frac{1}{2} + 0 * \frac{1}{4} + 1 * \frac{1}{8} + 1 * \frac{1}{16} = \frac{11}{16}$$

$$\begin{array}{cccc} 1/2\text{'s} & 1/4\text{'s} & 1/8\text{'s} & 1/16\text{'s} \\ 2^{-1} & 2^{-2} & 2^{-3} & 2^{-4} \end{array}$$

The problem with our loop is the 0.1. If you try to write $1/10$ as a *binary* fraction, it is infinitely repeating:

$$\frac{1}{10} = 0.00011001100110011\ldots$$

That means when 0.1 is represented in memory, there is a tiny error, because only so many bits can be stored—it has to cut off somewhere. You can see the error build up in the output of Example 2.5: there is fluctuation, but overall, the longer the loop runs, the worse the error gets.

Lesson: Use Inequalities with Floats

Because of this inherent limitation of floating-point values, testing floats for equality with either == or != is risky—values may not be exactly what you expect. Use inequalities (<, <=, >, >=) instead. A secondary lesson is to become aware of how much our intuition about fractions depends on base ten. We have a sense that $1/3$ is problematic because it is infinitely repeating in decimal, but we don't have the same intuition about $1/10$ in binary.

Even integers can behave unexpectedly in a program. In most languages, integers can **overflow** by becoming too large to store in the given number of bits. That is less of an issue in Python, though, because Python integers have unlimited precision.

Coda: Hexadecimal Numbers

When people want to write binary values, they often use base 16 (hexadecimal) instead, because it acts as a useful shorthand.

> **Base 16 (hexadecimal)** numbers use place values based on powers of 16 with digits 0–9 and A–F, where A=10, B=11, C=12, D=13, E=14, and F=15. For example,

$$\boxed{1} \quad \boxed{C} \quad \boxed{5} \quad = 1 * 256 + 12 * 16 + 5 * 1 = 453$$

$$\begin{array}{ccc} 256\text{'s} & 16\text{'s} & 1\text{'s} \\ 16^2 & 16^1 & 16^0 \end{array}$$

Base 16 is a good shorthand for binary because every four bits combine to form one hexadecimal digit. Table 2.9 lists built-in Python functions for converting integers between different bases.

TABLE 2.9   Base converters

`bin(n)`	
Return binary representation of integer n as a string beginning with `0b`.	

`hex(n)`	
Return hex representation of integer n as a string beginning with `0x`.	

`int(s, b)`	
Convert string s written in base b (default = 10) to an integer if possible.	

Example: Convert binary `0b10110010` to hexadecimal and decimal.

Convert to hexadecimal by splitting the bits into groups of four and interpreting each group separately:

binary	1011	0010
decimal	11	2
hex	B	2

so the result is `0xB2`. Convert to decimal by adding up the powers of either binary or hex; binary is $128 + 32 + 16 + 2 = 178$.

Example: Convert decimal 133 to binary and hex.

Convert to binary by first finding the highest power of 2 that will fit inside 133, which is 128. Then continue fitting the highest power with the remainder 5: a 4 will fit, and then a 1. This gives `0b10000101`. Convert binary to hex again in groups of four bits:

binary	1000	0101
hex	8	5

so the hex value is `0x85`.

EXERCISES

1. Determine the largest binary value that can be stored in one unsigned byte, along with its decimal and hexadecimal equivalents.

2. Determine the largest binary value that can be stored in two unsigned bytes, along with its decimal and hexadecimal equivalents.

3. Determine the largest unsigned integer that can be stored in a 32-bit machine.

4. Determine the largest unsigned integer that can be stored in a 64-bit machine.

5. Give the values of these Python expressions:

   (a) `hex(25)`

   (b) `bin(35)`

   (c) `int("0x1C", 16)`

   (d) `int("0b10101", 2)`

6. Give the values of these Python expressions:

   (a) `bin(178)`

   (b) `hex(157)`

   (c) `int("0b1111001", 2)`

   (d) `int("0xFE", 16)`

7. Show how each of these is stored as an unsigned integer in one byte. Write both binary and hexadecimal forms.

   (a) 87

   (b) 195

   (c) 18

   (d) 119

8. Show how each of these is stored as an unsigned integer in one byte. Write both binary and hexadecimal forms.

   (a) 93

   (b) 234

   (c) 203

   (d) 116

9. Convert these unsigned binary integers to decimal and hexadecimal:

   (a) 0b10010011

   (b) 0b00101101

   (c) 0b01001011

   (d) 0b10011110

10. Convert these unsigned binary integers to decimal and hexadecimal:

    (a) 0b01011100

    (b) 0b11000001

    (c) 0b10010111

    (d) 0b01001001

11. Convert these unsigned hexadecimal integers to binary and decimal:

    (a) 0x7D

    (b) 0xA1

    (c) 0x59

    (d) 0xBC

12. Convert these unsigned hexadecimal integers to binary and decimal:

    (a) 0xB6

    (b) 0x96

    (c) 0x04

    (d) 0xD9

13. Fix Example 2.5 so that it terminates.

14. Write a program to print a table of binary and hexadecimal representations of the decimal integers 1 through 100. You do not need any functions other than `main()`.

15. Convert the following decimal values to binary. Indicate which can or cannot be stored precisely as floats.

   (a) 0.25   (c) 0.3

   (b) 0.375   (d) 10.4375

16. Convert the following decimal values to binary. Indicate which can or cannot be stored precisely as floats.

   (a) 5.125   (c) 8.5

   (b) 14.6   (d) 3.65625

17. Convert the following binary values to decimal.

   (a) 0.1   (c) 0.111

   (b) 0.0101   (d) 10100.01

18. Convert the following binary values to decimal.

   (a) 111.1011   (c) 1000.10001

   (b) 0.0001   (d) 10111.11001

19. Determine whether or not the fraction $1/3$ can be stored exactly as a binary floating-point value.

20. Research the performance of Patriot missiles in the 1991 Gulf War.

21. Suppose you need to work with currency values in Python. Should you use ints or floats? Explain.

2.6   REPEATED INPUT

The next program, Example 2.6, illustrates a useful way to repeatedly ask the user for input. It uses the **random** library and provides a framework that we can develop into a more interesting program.

User Input Loop

Text-based programs often require input from the user until some ending condition occurs, such as the program reaching a successful conclusion or the user deciding to stop. It should seem natural to use a **while** loop to test the condition, but the question is, when should we ask for input?

In this setting, where the user is guessing a number, a guess has to be made before we can test whether or not it is correct. That suggests having an **input()** before the **while**. But if the guess is incorrect, we need to do something—at

```
1  # guess.py
2  # User guesses a random number.
3
4  from random import randint
5
6  def userguess(secret):
7      """Ask user for guesses until matching secret."""
8      guess = int(input("Your guess? "))
9      while guess != secret:
10         guess = int(input("Your guess? "))
11
12 def main():
13     secret = randint(1, 10)
14     userguess(secret)
15
16 main()
```

Example 2.6 Guess a number.

least respond to the user—and then ask for another guess. That means a second **input**() is needed at or near the bottom of the loop.

The full pattern looks like this:

```
response = input()
while response != finished:
    # handle the response
    response = input()
```

The condition in the loop may need to be different, but this represents the basic idea.

Spiral Development

Example 2.6 is not very exciting to run, but we can add features to it that make it more interesting. And that is a key aspect of software design: if you start with a simple, solid foundation, and then gradually add new features one at a time, you have a good chance of finishing with a reasonable product. Don't imagine that you need to be able to write out a long program straight through from the first line to the last—that's not how most people work. Instead, develop out from a working core.

This strategy of writing software outward from a working core is known as **spiral development**. To use it, follow these principles:

- Start as simple as possible, with a working program.

- Take one step forward at a time.

- Make sure the program works before taking another step.

In general, take small steps, because then it is easier to make sure you stay on a good path.

Refactoring

A "step" in program development can be adding a new feature or rewriting part of your code because now you can see a better way to do it. The development term for rewriting code to improve it without introducing new functionality is **refactoring**. Both adding new features and refactoring make progress toward creating a better program.

Example 2.6 could have been even simpler, but it is written this way to illustrate two common types of refactoring that you can begin using right away.

Example: Refactoring to introduce a new variable

> It might have been simpler to call `randint()` in `main()` without creating the variable `secret`. By naming the random integer "secret," anyone reading the code can tell what that value represents. This helps make code **self-documenting**, in that additional comments aren't necessary to explain what the code is doing. At the same time, we have saved the secret number for later use.

Example: Refactoring to introduce a new function

> It also would have been simpler to start with all of the code in `main()`. However, as existing functions start to get too long, look for ways to isolate coherent tasks that can be written as separate functions. If you find a good candidate, then look for the data that the function will need to do its work (its parameters), and what it needs to send back to the caller (its return value).

EXERCISES

1. Refactor this code to add a variable for the interest:

```
balance += rate * balance
```

2. Refactor this code to add a variable for the discriminant $b^2 - 4ac$:

```
root1 = (-b + sqrt(b**2 - 4*a*c))/(2*a)
root2 = (-b - sqrt(b**2 - 4*a*c))/(2*a)
```

3. Refactor this code to add a function for the opening greeting:

```
def main():
    print("Welcome to the Incredible Brain Machine")
    print("This program will tell your fortune")
    print("by asking you three simple questions.")
    print("Please answer truthfully")
    print("so that your fortune is accurate.")
    tell_fortune()
```

4. Refactor this code to add a function to calculate the present value:

```
pv = c/(1 + r)**t
```

5. Extend Example 2.6 by adding `print()` statements in `main()` to describe the game at the beginning and inform the user what happened at the end.

6. Extend Example 2.6 by adding `print()` statements in `userguess()` to inform the user when he or she has made an incorrect guess.

7. Extend Example 2.6 by counting the number of guesses made and returning the count to `main()`. Report the number of guesses to the user in `main()`.

8. Extend Example 2.6 to inform the user whether each incorrect guess was too high or too low.

9. Extend Example 2.6 to ask the user for the upper limit instead of always using 10.

10. After implementing the previous two exercises, can you reasonably guess the number if the upper limit is 1 million? Explain.

11. Write a program to repeatedly allow a user to calculate the area of circles. Stop when the input is -1; make sure the user knows that. Use appropriate functions and provide opening and closing messages. The area of a circle is πr^2.

12. Write a program to repeatedly allow a user to calculate the volume of spheres. Stop when the input is -1; make sure the user knows that. Use appropriate functions and provide opening and closing messages. The volume of a sphere is $\frac{4}{3}\pi r^3$.

13. Write a program to repeatedly allow a user to do mile-to-kilometer conversions. Stop when the input is -1; make sure the user knows that. Use appropriate functions and provide opening and closing messages. One mile is approximately 1.609 km.

14. Write a program to repeatedly allow a user to calculate maximum heart rate estimates, using the formula $208 - 0.7y$ for a person y years of age. Stop when the input is -1; make sure the user knows that. Use appropriate functions and provide opening and closing messages.

2.7   LISTS OF NUMBERS

Statistical questions involve collections of data rather than individual numbers in isolation. Python has a very powerful list mechanism for computing with lists of numbers. Example 2.7 shows how to calculate the sum of a Python list. It uses the same type of numeric accumulator we saw in Section 2.3, but with a somewhat new loop—a loop over a list. The function is named mysum() to avoid conflicting with the built-in **sum()**.

```
1  # mysum.py
2
3  def mysum(items):
4      """Return sum of values in items."""
5      total = 0
6      for item in items:
7          total += item
8      return total
9
10 def main():
11     data = [4, 9, 2, 8, 3, 2, 5, 4, 2]
12     print("My sum of", data, "is", mysum(data))
13
14 main()
```

Example 2.7   Sum.

Python Lists

A Python **list** stores an ordered sequence of items. That is, lists collect multiple items in a specific order, so that there is a first item, second item, third item, and so on. The **length** of a list is the number of items it contains. Lists are written inside square brackets " [" and "] ." A pair of empty brackets " [] " denotes the **empty list**.

Python lists are their own type, **list**. They may contain elements of any type, although most lists usually contain elements of the same type. Table 2.10 describes several of the built-in functions designed for lists. The **list()** function is a type converter like those in Table 2.4.

TABLE 2.10   List functions

`len(items)` Return the number of items in *items*.
`max(items)` Return the largest item in *items*.
`min(items)` Return the smallest item in *items*.
`sum(items)` Return the sum of items in *items*.
`sorted(items)` Return a new list of items in *items* in increasing order. For decreasing order, use `sorted(items, reverse=True)`.
`list(seq)` Return a list of items from the sequence *seq*.

Loop Over Items in a List

Python makes it easy to operate with each item in a list because lists are a type of sequence. Recall from Section 1.4 on page 19 that a **for** loop needs a *sequence* to loop over. Because lists are sequences, they do not require new syntax: to loop over the items in a list, just put the list in place as the sequence of a **for** loop:

```
for item in list:
    body
```

The *item* variable will take on the value of each item in the *list* inside the body of the loop. For example, the loop in line 6 of Example 2.7 loops over each of the items in its parameter `items`.

Creating Lists

There are three ways to create a list in Python:

1. Put elements inside square brackets, as in line 11 of Example 2.7:

   ```
   data = [4, 9, 2, 8, 3, 2, 5, 4, 2]
   ```

2. Convert a **range()** expression or other sequence to a list with the built-in **list()** function, as described earlier on page 20 in Section 1.4. For example,

   ```
   list(range(100)) = [0, 1, 2, ..., 99]
   ```

3. Write a **list comprehension** in this form:

```
[expression for variable in sequence]
```

This creates a list by evaluating the *expression* for each item in the *sequence*. List comprehensions offer the most flexibility of all three methods: they are like putting a little **for** loop inside square brackets.

Example: `[2*i for i in range(5)]` is `[0, 2, 4, 6, 8]`.

The expression `2*i` is evaluated for each $i = 0, 1, 2, 3, 4$ to create the items in the list.

Example: `[randint(1,1000) for _ in range(100)]`

This handy expression creates a list of 100 random integers between 1 and 1000. (Remember to import **randint**.) It is very useful for creating data to test list functions.

EXERCISES

1. Evaluate these Python expressions:

   (a) `list(range(6))`
   (b) `list(range(1, 20, 3))`
   (c) `[2*i for i in range(5)]`
   (d) `[n**2 for n in range(8)]`

2. Evaluate these Python expressions:

   (a) `list(range(2, 14))`
   (b) `list(range(10, 101, 10))`
   (c) `[2*i + 1 for i in range(15, 20)]`
   (d) `[m**3 for m in range(1, 5)]`

3. Write Python expressions to create these lists:

   (a) `[1, 2, 3, 4, 5, 6]`
   (b) `[2, 6, 10, 14, ..., 46]`
   (c) `[0, 0, 0, 0, ..., 0]` with fifty entries
   (d) A list of 1000 random integers either 0 or 1

4. Write Python expressions to create these lists:

   (a) `[11, 13, 15, 17, 19, 21, 23, 25]`
   (b) `[72, 69, 66, 63, ..., 3]`
   (c) `[0, 1, 2, 3, 4, 0, 1, 2, 3, 4, 0, ..., 4]` with fifty entries
   (d) A list of 1000 random integers between 1 and 6

5. Add a call to the built-in `sum()` in the `main()` of Example 2.7 to check the value returned by `mysum()`.

6. Modify Example 2.7 to run on a list of 500 random integers between -100 and 100. Compare the result of `mysum()` to the built-in `sum()`.

7. Write a `mean(items)` function to return the average of the values in the list *items*. Test your function on random data.

8. Write a `product(items)` function to return the product of the values in the list *items*. Test your function on random data.

9. Write a `count(target, items)` function to return the number of times *target* appears in the list *items*. Include a program to test your function.

10. Write a `count_evens(items)` function that returns the number of even integers in the list *items*. Test your function on random data.

11. Write a `count_odds(items)` function that returns the number of odd integers in the list *items*. Test your function on random data.

12. Write a `geometric_mean(items)` function to return the geometric mean of the values in the list *items*. The **geometric mean** of $x_1, x_2, \ldots, x_n$ is

$$(x_1 x_2 \ldots x_n)^{\frac{1}{n}} .$$

Test your function on random data.

13. Write a `sum_of_squares(items)` function to return the sum of the squares of the values in the list *items*. Test your function on random data.

14. Write a `rms(items)` function to return the root mean square of the values in the list *items*. The **root mean square** or **quadratic mean** of $x_1, x_2, \ldots, x_n$ is

$$\sqrt{\frac{x_1^2 + x_2^2 + \cdots + x_n^2}{n}} .$$

Test your function on random data.

15. Write a `variance(items)` function to return the variance of the population in the list *items*. The **population variance** of $x_1, x_2, \ldots, x_n$ is

$$\frac{(x_1 - m)^2 + (x_2 - m)^2 + \cdots + (x_n - m)^2}{n}$$

where m is the mean of $x_1, x_2, \ldots, x_n$. Test your function on random data.

16. Write a `std_dev(items)` function to return the standard deviation of a population in the list *items*. The **standard deviation** of a population is the square root of its population variance; see the previous exercise. Test your function on random data.

17. Write a `sample_variance(items)` function to return the sample variance of the list *items*. The **sample variance** of $x_1, x_2, \ldots, x_n$ is

$$\frac{(x_1 - m)^2 + (x_2 - m)^2 + \cdots + (x_n - m)^2}{n - 1}$$

where m is the mean of $x_1, x_2, \ldots, x_n$. Test your function on random data.

18. Write a `sample_std_dev(items)` function to return the standard deviation of a sample in the list *items*. The **sample standard deviation** is the square root of its sample variance; see the previous exercise. Test your function on random data.

19. Write a function `randints(a, b, n)` to return a list of n random integers between a and b (inclusive). Use your function to print a list of fifty random integers between 1 and 10.

20. Write a function `randfloats(a, b, n)` to return a list of n random floating-point values between a and b (inclusive). Use your function to print a list of fifty random floats between 0 and 1.

2.8 LIST INDEXING AND SLICING

While a **for** statement can loop over each item in a list, there are times when we need to access specific elements or sections of lists. Python provides powerful indexing and slicing mechanisms for this purpose, illustrated in the `mymin()` function of Example 2.8. This function is similar to `mymax2()` from Example 2.1: `mymin()` begins by assuming the first element is the smallest, and then checks the smallest seen so far against each of the remaining items.

```
1  def mymin(items):
2      """Return smallest element in items."""
3      smallest = items[0]
4      for item in items[1:]:
5          if item < smallest:
6              smallest = item
7      return smallest
```

Example 2.8 Minimum.

Picturing Lists

Python lists are **containers**, meaning they store references to other objects. Think of a **reference** as a name or label that points to the object, as drawn for assignment on page 45. References of objects in lists are stored in consecutive memory locations, numbered starting at 0. Each number is called the **index** (plural indices) of the object at that location. Thus, the list [10, 3, -1, 8, 24] is stored as in Fig. 2.3, and the item at index 2 is −1.

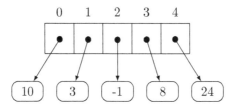

Figure 2.3 List indices and references to contents.

⟶ Note: The indices for a list of 5 items are the same as `range(5)`.

In fact, this is how all **sequence** types store data, so you will see other types later that work in the same way. Negative indices may also be used to count back from the end of a sequence, as shown in Fig. 2.4.

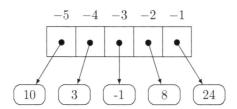

Figure 2.4 Negative indices.

Indexing and Slicing

Python provides **indexing** to access one item in a list, as well as a powerful **slicing** operation that allows accessing groups of items together. These operations are listed in Table 2.11. Slices work like `range()`, stopping before their second parameter and having an optional step size.

⟶ Note: Slices are themselves lists, but indexing returns a single object.

⟹ Caution: The slice `items[i:j]` does not include `items[j]`.

TABLE 2.11   Indexing and slicing

---

items[*i*]
Object in *items* at index *i*. If *i* < 0, count back from end. Raises
`IndexError` if *i* is not a valid index.

---

items[*i*:*j*]
Slice [*items*[*i*], *items*[*i* + 1], ..., *items*[*j* − 1]]. If *i* is omitted,
start at the beginning; if *j* is omitted or *j* > `len(s)`, go to the end.

---

items[*i*:*j*:*k*]
Slice with step size *k*, [*items*[*i*],*items*[*i* + *k*],...], stopping before
index *j*. If *k* < 0, step backward, still starting at *i* and stopping before *j*.
If *i* is omitted with negative *k*, the "beginning" is index −1, and if *j* is
omitted, the "end" is index 0.

Example: If `items` = `[10, 3, -1, 8, 24]` then

- `items[1]` is 3,

- `items[-1]` is 24, and

- `items[2:4]` is the list `[-1, 8]`.

Slice Shorthands

As noted in Table 2.11, slices have useful shorthands when one or both of the
parameters are omitted.

Example: `items[3:]` slices from index 3 to the end.

Example: `items[:10]` slices from the beginning to index 9.

Example: `items[:]` is a **full slice**, creating a copy of `items`.

Modifying Lists

Indexing and slicing can be used to modify lists by assigning new values to
either an indexed location or slice. There is no new syntax, so the following
examples illustrate how it works.

Example: If `items` = `[-2, 5, 3, -8, 7]` then

- `items[0]` = 10 changes `items` to `[10, 5, 3, -8, 7]`,

- `items[1:3]` = `[0]` changes the original `items` to `[-2, 0, -8, 7]`, and

- `items[1:3]` = 0 is not allowed because only a list (or other sequence)
  may be assigned to a slice.

⟹ Caution: Never modify a list that is driving a **for** loop from inside the loop.

Indexing may only be used to access elements already in a list; indexing cannot be used to add a new entry to a list.

Example: If items = [0, 1, 2], then attempting to set items[3] = 3 will generate an "index out of range" IndexError.

EXERCISES

1. Suppose nums = [35, 26, 19, 23, 6, 8, 18, 1, 34]. Give the value of each of these expressions:

   (a) nums[3]   (e) nums[2:5]

   (b) nums[-1]   (f) nums[3:7]

   (c) nums[:4]   (g) nums[1::2]

   (d) nums[6:]   (h) nums[::-1]

2. Suppose nums = [32, 18, 25, 34, 23, 24, 40, 37, 4, 31]. Give the value of each of these expressions:

   (a) nums[2]   (e) nums[1:4]

   (b) nums[-1]   (f) nums[5:8]

   (c) nums[:-1]   (g) nums[::2]

   (d) nums[3:]   (h) nums[::-1]

3. Suppose items is a nonempty list. Write Python expressions for:

   (a) The first element (counting from the beginning)

   (b) The last element

4. Suppose items is a list with at least three elements. Write Python expressions for:

   (a) The third element (counting from the beginning)

   (b) The next-to-last element

5. Let items = [7, 4, 27, 26, 2, 38, 19, 10, 34]. Write expressions using items to produce each of these values:

   (a) 38   (c) [2, 38, 19]

   (b) [7, 4]   (d) [4, 26, 38, 10]

6. Let items = [11, 29, 22, 36, 40, 25, 14, 7, 16]. Write expressions using items to produce each of these values:

    (a) [29]

    (b) [14, 7, 16]

    (c) [36, 40, 25, 14]

    (d) [16, 7, 14, 25, 40]

7. Suppose nums = [49, 41, 3, 46, 9, 12, 20]. Show the result of each of these assignments, starting over with the original value of nums for each.

    (a) nums[3] = 6

    (b) nums[0] += 10

    (c) nums[:2] = [0, 1, 2]

8. Suppose nums = [16, 40, 37, 32, 29, 47, 26, 50]. Show the result of each of these assignments, starting over with the original value of nums for each.

    (a) nums[-1] = 10

    (b) nums[2] += nums[1]

    (c) nums[1:4] = []

9. Let items = [34, 35, 13, 2, 29, 38, 7, 38, 5, 22]. Write one assignment statement (each) that would change items to:

    (a) [34, 35, 13, 2, 29, 38, 38, 38, 5, 22]

    (b) [34, 35, 36, 37, 38, 7, 38, 5, 22]

    (c) [34, 35, 13, 2, 29, 38, 7, 38, 39, 40, 41]

10. Let items = [5, 23, 17, 36, 3, 20, 31, 23, 2, 36]. Write one assignment statement (each) that would change items to:

    (a) [5, 23, 0, 36, 3, 20, 31, 23, 2, 36]

    (b) [5, 23, 17, 36, 36]

    (c) [5, 23, 17, 36, 3, 4, 5, 6, 7, 20, 31, 23, 2, 36]

11. Incorporate mymin() from Example 2.8 into a complete program that tests mymin() against the built-in min() on random data.

12. Write a mymax(items) function that returns the largest item in the list *items* without using the built-in max() function. Test your function on a random list of numbers and compare it against max().

13. Write a first(items) function to return the first item in the list *items*. Include a program to test your function.

14. Write a last(items) function to return the last item in the list *items*. Include a program to test your function.

15. Write a `rest(items)` function to return a list containing all but the first item in the list *items*. Include a program to test your function.

16. Write a function `myreversed(items)` to return a list containing the items in *items* in reverse order, meaning opposite of their original order. Do not use the built-in `reversed()` function or modify the original list. Test your function on random lists.

17. Write a `mychoice(items)` function to return a random element from the list *items*. Do not use the `choice()` function from the `random` module, but you may use other functions from Table 1.8. Include a program to test your function.

18. Write a `median(items)` function to return the median of the values in the list *items*. To find the **median**, sort the values in increasing order, and then if the list has odd length, the median is the middle element; otherwise, if it has even length, the median is the average of the two middle values. Do not modify the original list, and test your function on random lists.

19. Write a `swap(items, i, j)` function to swap *items*[i] with *items*[j]. Write a program to test your function.

20. (Requires Exercise 19.) Write a function `myshuffle(items)` that shuffles the list *items* without using the `shuffle()` function from the `random` module. Use the `swap()` function from the previous exercise and this outline:

> for every index i:
>   pick a random index j ≥ i
>   swap items i and j

Include a program to test your function.

2.9 LIST ACCUMULATION

Many problems require building a list one item at a time within a loop. This process is analogous to numeric accumulation, and the Python code reflects their close similarity. The `evens()` function in Example 2.9 accumulates a list of all the even integers from any given list. Compare it closely with `mysum()` from Example 2.7: both use a loop over a list, but whereas `mysum()` has a numeric accumulator (accumulating with addition), `evens()` uses a list accumulator with a new operation, concatenation.

Combining Lists: Concatenation

Concatenation is the operation of putting two lists together, one after the other, to create one new list. Because this can be thought of as "adding"

```
1  def evens(items):
2      """Return list of even values in items."""
3      result = []
4      for item in items:
5          if item % 2 == 0:
6              result += [item]
7      return result
```

Example 2.9 Even values in a list.

two lists, Python uses the addition symbol to represent concatenation; see Table 2.12. Multiplication is repeated addition, and so there is also a repeated concatenation operator.

TABLE 2.12   List concatenation

x + y List consisting of elements of x followed by elements of y.
x += y Append list y onto x. Similar to x = x + y.
n * x Repeatedly concatenate list x with itself n times.
x * n Same as n * x.

Example: [1, 4, 7, 8] + [9, 2] is [1, 4, 7, 8, 9, 2].

Example: 3*[0, 1] is [0, 1, 0, 1, 0, 1].

Example: items[:i] + items[i:] is a copy of items.

> The two pieces make up the whole list items. This expression helps explain why Python slices work the way they do.

Example: items += [100]

> **Appends** the value 100 to the end of list items.

Example: items = [25] + items

> **Prepends** item 25 to the front of list items. The concatenation shorthand += will only append because it is short for adding on the right.

⟹ Caution: Concatenation requires lists—attempting to append or prepend an individual item with concatenation will not work. More generally, concatenation only works between objects of the same type.

List Accumulators

List accumulation usually involves appending to an accumulator, and so +=
may be used in the same way as for a numeric accumulator:

```
accumulator = []
loop:
    accumulator += list to append
```

The accumulator could start with something other than an empty list, but
that is the most common.

Example:

```
result = []
for i in range(4):        output    0 [0]
    result += [i**2]      ------->   1 [0, 1]
    print(i, result)                 2 [0, 1, 4]
                                     3 [0, 1, 4, 9]
```

The variable i gets its value from the **range()**, while result
appends i^2 each time the loop runs.

Types Determine Operations

Understanding Python types is important, because an object's type determines
the operations it supports (see page 46). For example, now that we have two
different meanings for "+"—addition and concatenation—you will need to
watch object types carefully to decide which operation is being used.

Example: Let items = [8, 2, 1, 3, 5, 4]. Then

- items[2] + items[4] is $1 + 5 = 6$, and

- items[:2] + items[4:] is [8, 2, 5, 4].

List Comprehensions with Conditions

It is possible to add **if** conditions to list comprehensions (see page 76), giving
another way to write the evens() function of Example 2.9. The condition is
simply added to the end of the comprehension:

```
[expression for variable in sequence if condition]
```

This construction creates a list by evaluating *expression* for each value of
variable in *sequence* for which *condition* is true.

Each **for** or **if** in a comprehension introduces a **clause**, and any number of clauses may be put after the initial **for**:

$$\underbrace{\textit{expression}\ \texttt{for variable in sequence}}\ \underbrace{\texttt{if condition}}$$

Example 2.10 shows how to write **evens()** using this technique. It helps to see the simpler expression

```
[item for item in items]
```

as simply copying the list **items**. Then the **if** clause selects only the even items. This is a terse—even cryptic—style, but it is powerful once you get used to it.

```
1  def evens(items):
2      """Return list of even values in items."""
3      return [item for item in items if item % 2 == 0]
```

Example 2.10 Even values using list comprehension.

EXERCISES

1. Suppose a = [3, 15, 30, 19, 5, 10] and b = [32, 39, 35, 27]. Determine the value of each of these expressions:

    (a) a + b
    (b) b[1:] + a
    (c) 3 * b

2. Suppose a = [47, 42, 44, 12] and b = [40, 38, 2, 32, 3]. Determine the value of each of these expressions:

    (a) b + 2*a
    (b) a[:3] + b[2:]
    (c) a[3] + b[2]

3. Suppose a = [3, 15, 30, 19, 5, 10] and b = [32, 39, 35, 27].

    (a) Determine the effect of a += b.
    (b) Determine the effect of a = [b[-1]] + a.

4. Suppose a = [47, 42, 44, 12] and b = [40, 38, 2, 32, 3].

    (a) Determine the effect of b = a[:2] + b.
    (b) Determine the effect of b += b.

5. Suppose a and b are lists.

   (a) Write an expression to concatenate the first three elements of a with the last four elements of b.

   (b) Write an expression to concatenate a with the first element of b.

6. Suppose a and b are lists.

   (a) Write an expression to concatenate all but the first element of a with the first two elements of b.

   (b) Write an expression to append the last element of a onto b.

7. Write a Python statement to set counts to name a list of 25 0's.

8. Write a Python statement to append 15 0's to the end of the list counts.

9. Determine the output of this code:

```
result = []
for i in range(4):
    result += [2*i + 1]
    print(i, result)
```

10. Determine the output of this code:

```
result = []
for i in range(5):
    print(i, result)
    result += [25 - 3*i]
```

11. Determine the value of: `[i**2 for i in range(10) if i%2 == 1]`

12. Determine the value of: `[2**i for i in range(10) if i%3 == 1]`

13. Incorporate evens() from Example 2.9 into a complete program that tests it on random data.

14. Incorporate evens() from Example 2.10 into a complete program that tests it on random data.

15. Write an odds(items) function to return a list of only the odd integers from the list items. Test your function on random lists.

16. Write a function myreversed(items) to return a list containing the items in items in reverse order using a list accumulator. Do not use the built-in reversed() function or modify the original list. Test your function on random lists.

17. Write a function rotate_right(*items*) to return a list containing the elements in *items* shifted one place to the right, wrapping the last element around to the front. For example,

$$[0, 1, 2, 3] \xrightarrow{\text{rotate right}} [3, 0, 1, 2]$$

Do not modify the original list. Include a program to test your function.

18. Write a function rotate_left(*items*) to return a list containing the elements in *items* shifted one place to the left, wrapping the first element around to the end of the list. For example,

$$[0, 1, 2, 3] \xrightarrow{\text{rotate left}} [1, 2, 3, 0]$$

Do not modify the original list. Include a program to test your function.

19. Write a function randints(a, b, n) to return a list of n random integers between a and b (inclusive) using a list accumulator. Use your function to print a list of fifty random integers between 1 and 10.

20. Write a function randfloats(a, b, n) to return a list of n random floating-point values between a and b (inclusive) using a list accumulator. Use your function to print a list of fifty random floats between 0 and 1.

21. Write a function divisors(n) to return a list of the proper positive divisors of the integer n, including 1 but excluding n itself. Use your function to print a table of values of the divisors of 2 through 20.

22. (Requires Exercise 21.) Write a function isperfect(n) to return True if n is a **perfect** number, meaning it is positive and equal to the sum of its proper divisors. Use the previous exercise, and write a program to find all perfect numbers less than 10,000.

2.10   SEARCHING A LIST

Many computational tasks require searching a list, either to determine where an item is or simply whether or not it is present. Example 2.11 uses a **brute-force** strategy to solve this problem, looking through the list one slot at a time, until the item is found. The find() function uses an early return if the item is found (see page 46); otherwise, if the item is not present, the loop ends and −1 is returned. This approach is called **linear search**.

```
1  def find(target, items):
2      """Return first index of target in items or -1."""
3      for i in range(len(items)):
4          if target == items[i]:
5              return i
6      return -1
```

Example 2.11 Find item in list.

Indexed Loops

In order to return where the item is located, `find()` in Example 2.11 uses a loop over the list's indices rather than a loop over its elements. There is no special syntax: it is simply a conscious choice to make when writing loops with lists. If you need the index for any reason, use a loop over indices:

```
for i in range(len(items)):
```

If you do not need the index, then a loop over elements (page 75) is usually simpler:

```
for item in list:
```

A loop over elements like this could be used to decide whether or not a list contains an item, but not where it is.

Boolean Searches

To answer the simpler question of whether or not a list contains a particular element, Python has built-in **in** and **not in** operations, given in Table 2.13. Because they return boolean values, these operators are most often used in **if** statements or **while** loops.

TABLE 2.13 List containment

`item in list`
True if *item* is equal to an element of *list*; otherwise, **False**.
`item not in list`
True if *item* is not equal to any element of *list*; otherwise, **False**.

Searching Sorted Lists

In a sorted list, there is a better strategy than brute force. **Binary search** is the name of the strategy people use for guessing a number when told whether each guess is too high or too low: keep track of the range in which the number is trapped, and make the next guess in the middle of that range. This is known as a **divide-and-conquer** or decrease-and-conquer strategy, because the list is repeatedly divided in half during the search.

Applied to Python lists, this strategy keeps track of *left* and *right* indices that trap where the number must be located. At each step, the index *mid* is calculated as halfway between left and right, and the element there is tested.

Example: Binary search for 31 in [2, 17, 19, 23, 31, 40, 48]

> Table 2.14 tracks the *left*, *right*, and *mid* indices used for this search. Notice that after the first comparison, *left* can be set to 4 because 31 cannot be anywhere with index ≤ 3.

TABLE 2.14 Binary search example

left	*right*	*mid*	**result**
0	6	3	too low, so update *left*
4	6	5	too high, so update *right*
4	4	4	found it, return 4

The indices *left* and *right* will cross if the item is not found, and the process can stop.

EXERCISES

1. Fill in a table like Table 2.14 showing the indices used in a binary search for 31 in the list [7, 13, 14, 21, 22, 28, 31, 37].

2. Fill in a table like Table 2.14 showing the indices used in a binary search for 7 in the list [7, 13, 14, 21, 22, 28, 31, 37].

3. Fill in a table like Table 2.14 showing the indices used in a binary search for 20 in the list [7, 13, 14, 21, 22, 28, 31, 37].

4. Fill in a table like Table 2.14 showing the indices used in a binary search for 40 in the list [7, 13, 14, 21, 22, 28, 31, 37].

5. Incorporate `find()` from Example 2.11 into a complete program that tests it on random data.

6. Rewrite `find()` from Example 2.11 to use a **while** loop instead of **for**. Include a program to test your function.

7. Write a `contains(target, items)` function to return `True` if the list *items* contains *target* and `False` otherwise. Do not use **in** or **not in**. Include a program to test your function.

8. Write a `find_last(target, items)` function to return the index of the last occurrence of *target* in *items*, or −1 if it is not found. Include a program to test your function.

9. Write a program that uses binary search to guess a number between 1 and 100 chosen by the user. After each guess, the user indicates if the guess was too high, too low, or correct. You do not need to write any functions other than `main()`.

10. Write a `binary_search(target, items)` function that uses binary search to return the index where *target* is located in the sorted list *items* or −1 if the item is not found. Include a program to test your function. Do not worry about duplicate items.

11. You might have expected "+" to arithmetically add the elements of two lists, as long as they were lists of numbers. Write an `sumlists(a, b)` function to return a list of the sums of corresponding items in a and b (`a[0] + b[0]`, etc.). If one list is shorter, treat missing items as 0. Test your function on random data.

12. Write a `fibonacci(n)` function to return a list of the first *n* Fibonacci numbers. The **Fibonacci numbers** start with two 1's and then each next element is the sum of the two previous values:

$$1, 1, 2, 3, 5, 8, 13, 21, 34, 55, \ldots$$

Include a program to test your function.

2.11   RECURSION

Rather than using **while** or **for** loops, some programs implement repetition through recursion. A function is **recursive** if it calls itself, usually on a smaller argument. The sequence of calls must eventually terminate in a **base case** that does not generate a new recursive call.

Example 2.12 shows how to compute factorials recursively. The **factorial** of n for integer $n \geq 0$, is

$$n! = n \cdot (n - 1) \cdot (n - 2) \cdots 3 \cdot 2 \cdot 1.$$

For example, $4! = 4 \cdot 3 \cdot 2 \cdot 1 = 24$.

A precise mathematical definition (avoiding "...") uses cases:

$$n! = \begin{cases} 1 & \text{if } n = 0 \\ n \cdot (n - 1)! & \text{if } n > 0. \end{cases}$$

Compare this with the code in Example 2.12; they are nearly identical.

```
1  def factorial(n):
2      """Return n! = 1*2*3*...*n."""
3      if n <= 1:
4          return 1
5      return n * factorial(n - 1)
```

Example 2.12 Recursive factorial.

Tracing Recursive Calls

To understand how the function in Example 2.12 works, trace what it does when n = 4. Because 4 is not less than or equal to 1, the if condition is false, and we have:

$$\text{factorial(4)} \quad \text{returns} \quad 4 * \text{factorial(3)}$$

To return this, factorial(3) must be evaluated, and that follows the same steps, but this time with n = 3. The if condition is false, and so factorial(3) returns 3 * factorial(2), which we can incorporate into the whole computation like this:

```
factorial(4)   returns   4 * factorial(3)
               =         4 * (3 * factorial(2))
```

The next call to factorial(2) returns 2 * factorial(1):

```
factorial(4)   returns   4 * factorial(3)
               =         4 * (3 * factorial(2))
               =         4 * (3 * (2 * factorial(1)))
```

This seems like it could go on forever! But that's the key: it doesn't go on forever. At the next call, the if condition is true, and factorial(1) returns 1. No new call is made, and the computation winds its way back to return 24:

```
factorial(4)   returns   4 * factorial(3)
               =         4 * (3 * factorial(2))
               =         4 * (3 * (2 * factorial(1)))
               =         4 * (3 * (2 * 1))
               =         4 * (3 * 2)
               =         4 * 6
               =         24
```

Requirements

From this example, we can see two general requirements for recursive functions:

Base cases need to be checked first, before calling the function recursively.

Recursive calls must be on *smaller* arguments; otherwise, the base case(s) will never be reached.

These two requirements together ensure that the recursion stops and a value is returned. Otherwise, a `RecursionError` will be raised with a "maximum recursion depth exceeded" message.

In fact, there is a limit to what the interpreter can compute recursively, because it needs to keep track of all of the ongoing, partial computations. Calling `factorial()` with a large n will result in the same error.

EXERCISES

1. Trace the computation of add$(6, 3)$ using this recursive definition of addition for integer $n \geq 0$:

$$\text{add}(m, n) = \begin{cases} m & \text{if } n = 0 \\ 1 + \text{add}(m, n - 1) & \text{if } n > 0 \end{cases}$$

2. Write a recursive Python function `add(m, n)` to compute the add() function in the previous exercise. Test your function against regular addition.

3. Trace the computation of mul$(5, 4)$ using this recursive definition of multiplication for integer $n \geq 0$:

$$\text{mul}(m, n) = \begin{cases} 0 & \text{if } n = 0 \\ m + \text{mul}(m, n - 1) & \text{if } n > 0 \end{cases}$$

4. Write a recursive Python function `mul(m, n)` to compute the mul() function in the previous exercise. Test your function against regular multiplication.

5. Trace the computation of power$(2, 5)$ using this recursive definition of exponentiation for $m > 0$, integer $n \geq 0$:

$$\text{power}(m, n) = \begin{cases} 1 & \text{if } n = 0 \\ m \cdot \text{power}(m, n - 1) & \text{if } n > 0 \end{cases}$$

6. Write a recursive Python function `power(m, n)` to compute the power() function in the previous exercise. Test your function against the built-in `pow()` function.

7. Write a recursive Python function `fib(n)` to return the nth **Fibonacci number**, where each term is the sum of the two previous values:

$$1, 1, 2, 3, 5, 8, 13, 21, 34, 55, \ldots$$

For example, `fib(1)` = 1 and `fib(4)` = 3. Include a program to test your function.

8. Compare the `fib()` function of the previous exercise to the `fibonacci()` function of Exercise 12 in Section 2.10. In particular, compare running `fib(100)` and `fibonacci(100)`. Try to explain any differences you observe.

9. Trace the computation of gcd(48, 40) using this recursive definition of the **GCD** (greatest common divisor) of integers m, $n \geq 0$:

$$\gcd(m, n) = \begin{cases} m & \text{if } n = 0 \\ \gcd(n, m \bmod n) & \text{if } n > 0 \end{cases}$$

Notice that $m \bmod n$ is always less than n.

10. Write a recursive Python function `gcd(m, n)` to compute the gcd() function in the previous exercise. Use the remainder operator `%` to compute $m \bmod n$. Include a program to test your function.

11. Python lists are naturally recursive: they can get gradually smaller with a slice that removes one element, and the base case is an empty list. Write a recursive `length(items)` function to return the length of the list *items* without using the built-in `len()` function. Test your function against `len()`.

12. Write a recursive `mymin(items)` function to return the smallest item in the nonempty list *items* without using the built-in `min()` function. Test your function against `min()`. Hint: the base case for a nonempty list is a list with one element.

CHAPTER 3

Text

3.1 STRINGS

Python stores text in the string data type, **str**. Processing text is quite different from computing with numbers, but Python strings share many similarities with lists. For example, strings are often taken apart and put back together in different ways. Example 3.1 illustrates this by forming the present indicative conjugations of a family of regular Spanish verbs.

```python
# conjugator.py

def conjugate(verb):
    """Return conjugation of regular -ar Spanish verbs."""
    stem = verb[:-2]
    return [stem + "o", stem + "as", stem + "a",
            stem + "amos", stem + "áis", stem + "an"]

def main():
    verb = input("Enter an -ar verb: ")
    print("Present indicative conjugation of " + verb + ":")
    for form in conjugate(verb):
        print(form)

main()
```

Example 3.1 Verb conjugator.

Strings Are Sequences

Python **sequence types** share many common features because they represent sequences of objects, where there is a first, second, third, etc. Strings are the third sequence type we have seen: **range**() returns a sequence, lists are sequences, and so are strings. A fourth sequence type, tuple, will be introduced later in Section 4.1.

All sequence types are indexed with indices starting at 0. Strings are sequences of characters stored in consecutive memory locations; for example, the string "hello" can be pictured as in Fig. 3.1.

$$\begin{array}{ccccc} 0 & 1 & 2 & 3 & 4 \\ \boxed{\text{h}} & \boxed{\text{e}} & \boxed{\text{l}} & \boxed{\text{l}} & \boxed{\text{o}} \\ -5 & -4 & -3 & -2 & -1 \end{array}$$

Figure 3.1   Strings in memory.

Table 3.1 summarizes operations common to all sequence types, including strings. These operations were described earlier for lists in Sections 2.7–2.10.

Subtleties

While all of the operations in Table 3.1 apply to strings, there are a few subtleties to be aware of:

- There is no separate character type, so indexing (or slicing) a string returns a string.

- The **min**() and **max**() functions find the smallest and largest characters, so they also return strings.

- The **in** and **not in** operations test whether or not x is a **substring** of s, meaning that it occurs as a consecutive piece of s.

Example: "bcd" **in** "abcde" is true.

   "bcd" is a substring of "abcde".

Example: "ad" **not in** "abcde" is true.

   "ad" is not a substring of "abcde", because it does not occur as a consecutive slice.

Example: min("python") is "h".

Accented Characters

Accented characters may be typed directly into Python programs, as on line 7, but you can also add an acute accent (as in "á") to a character within a program by concatenating the accent's code "\u0301" after the character.

Example: "a" + "\u0301" or "a\u0301" produces "á."

We will learn more about these codes in Section 3.3.

TABLE 3.1   Sequence operations

`s[i]`
Object in *s* at index *i*. If $i < 0$, count back from end. Raises `IndexError` if *i* is not a valid index.

`s[i:j]`
Slice `s[i]`, `s[i + 1]`, ..., `s[j - 1]`. If *i* is omitted, start at the beginning; if *j* is omitted or $j > \text{len}(s)$, go to the end.

`s[i:j:k]`
Slice `s[i]`,`s[i + k]`,... stopping before `s[j]`. If $k < 0$, step backward, still starting at *i* and stopping before *j*. If *i* is omitted with negative *k*, the "beginning" is index -1, and if *j* is omitted, the "end" is index 0.

`s + t`
Concatenate *s* with *t*.

`s += t`
Append *t* onto *s*. Similar to `s = s + t`.

`n * s` or `s * n`
Repeatedly concatenate *s* with itself *n* times.

`len(s)`
Return the length of *s*.

`max(s)`
Return the largest item in *s*.

`min(s)`
Return the smallest item in *s*.

`x in s`
True if *x* is contained in *s*; otherwise, `False`.

`x not in s`
True if *x* is not contained in *s*; otherwise, `False`.

Line Continuation

Line continuation is used in Example 3.1 from line 6 to line 7 to break an extra-long line. It is allowed at that point because the line is broken after a comma in a list. Not all lines of code may be split like this, but it is safe to do so after commas inside parentheses, square brackets, or braces.

⟶ Note: The continuation line (line 7) is indented to be just inside the opening bracket of the list, aligned with the first item in the list.

Be Aware of Types

Recall from Section 2.1 that the type of a piece of data in a Python program determines which operations can be used on it. Text is stored in strings, whereas numeric data is either integer or float. Table 2.4 on page 47 lists type converting functions if you need to change from one type to another.

EXERCISES

1. Suppose word = "albatross". Give the value of each of these expressions:

    (a) word[3] (c) word[2:5]

    (b) word[:4] (d) word[::2]

2. Suppose word = "palatable". Give the value of each of these expressions:

    (a) word[-1] (c) word[1:6:2]

    (b) word[4:] (d) word[::-1]

3. Suppose word = "rehearse". Give the value of each of these expressions:

    (a) word[5] (c) word[3:6]

    (b) word[:-1] (d) word[1::2]

4. Suppose word = "shellfish". Give the value of each of these expressions:

    (a) word[-1] (c) word[2:5]

    (b) word[5:] (d) word[::-1]

5. Determine the value of these expressions:

    (a) "event" in "seventies" (b) "seed" in "squelched"

6. Determine the value of these expressions:

    (a) "dome" in "seldom errs" (b) "pig" in "pigeonhole"

7. Let `phrase` = `"string quartet in d minor"`. Write expressions using `phrase` to produce each of these values:

    (a) `"string"`   (b) `"quart"`   (c) `"minor"`   (d) `"ring"`

8. Let `phrase` = `"delivery service"`. Write expressions using `phrase` to produce each of these values:

    (a) `"deli"`   (b) `"very"`   (c) `"ice"`   (d) `"reviled"`

9. Let `phrase` = `"twenty-four hours"`. Write expressions using `phrase` to produce each of these values:

    (a) `"twenty"`   (b) `"four"`   (c) `"hour"`   (d) `"newt"`

10. Let `phrase` = `"salami sandwich"`. Write expressions using `phrase` to produce each of these values:

    (a) `"sand"`   (b) `"ami"`   (c) `"ich"`   (d) `"alas"`

11. Write a boolean expression that is `True` if the first character of the string `word` is a vowel (not including "y").

12. Write a boolean expression that is `True` if the second character of the string `word` is a vowel (including "y").

13. Modify Example 3.1 to instead conjugate regular Spanish "-er" verbs.

14. Modify Example 3.1 to instead conjugate regular Spanish "-ir" verbs.

15. Modify Example 3.1 to return the corresponding pronoun(s) with each form.

16. Modify the `conjugate()` function of Example 3.1 to conjugate regular Spanish "-ar" and "-er" verbs. Isolate the vowel in the verb that is different and use it to create the correct endings.

17. Write a conjugation function for a language and regular verb family of your choice. Include a program to test your function.

18. Write a `username(first, last)` function to return a computer system username given a person's first and last names. The username is made up of at most 12 characters of the last name, followed by the first initial of the first name, followed by a "1." Include a program to test your function.

19. Write a `pig_latin(word)` function to translate *word* into Pig Latin, assuming *word* either starts with a vowel or only one consonant. Include a program to test your function.

20. Write a `pig_latin(word)` function to translate *word* into Pig Latin without the limitation of the previous exercise. Include a program to test your function.

21. Write an adverb(*adjective*) function to return the adverb form of *adjective* in English. Do not worry about irregularities, but try to capture some of the regular forms. Include a program to test your function.

22. Write a plural(*noun*) function to return the plural form of *noun* in English. Do not worry about irregularities, but try to capture some of the regular forms. Include a program to test your function.

23. Like lists, Python strings are naturally recursive: they can get gradually smaller with a slice that removes one character, and the base case is an empty string. Write a recursive length(s) function to return the length of the string *s* without using the built-in len() function. Test your function against len().

3.2   STRING ACCUMULATION

Strings are useful for capturing a wide variety of information. For example, genetic sequences can be represented by character strings. Example 3.2 generates a random string of DNA and displays it with its complement, using string accumulation to build both the random sequence and its complement.

A Brief Introduction to DNA

DNA is a double helix of two chains of nucleotides. Each nucleotide base can be represented by a single letter, and so a chain of nucleotides can be thought of as a string. Even though DNA has two chains (also called strands), the two are closely related: given one, it is easy to calculate the other. Thus, DNA is generally described as a string of characters representing the nucleotide bases of one of its strands.

The four bases that make up DNA are adenine (A), cytosine (C), guanine (G), and thymine (T). They occur in complementary pairs, A ↔ T and C ↔ G. The **complement** of a strand of DNA swaps each base with its complementary base.

Example: The complement of AGGTC is TCCAG.

It is the complement that forms the second strand in DNA because complementary pairs bind across from each other in base pairs to form the double helix. Thus, if AGGTC occurs in a strand of DNA, it is linked to its complement like this:

```
 1  # dna.py
 2
 3  from random import choice
 4
 5  def complementary_base(base):
 6      """Return complement of single base."""
 7      if base == "A":
 8          return "T"
 9      elif base == "T":
10          return "A"
11      elif base == "C":
12          return "G"
13      elif base == "G":
14          return "C"
15      return base    # leave anything else
16
17  def complement(dna):
18      """Return complement of dna strand."""
19      result = ""
20      for base in dna:
21          result += complementary_base(base)
22      return result
23
24  def random_dna(length=30):
25      """Return random strand of dna of given length."""
26      fragment = ""
27      for _ in range(length):
28          fragment += choice("ACGT")
29      return fragment
30
31  def main():
32      dna = random_dna()
33      print("Sequence   :", dna)
34      print("Complement:", complement(dna))
35
36  main()
```

Example 3.2 DNA sequences.

These strands are directional, so that if the top strand reads from left to right, then the bottom strand reads from right to left. Thus, the **reverse complement** of a strand is also important.

Example: The reverse complement of AGGTC is GACCT.

String Concatenation

As outlined in Table 3.1 in the last section, strings concatenate with "+" in the same way as lists. String concatenation allows for much more flexible output in print() statements. For example, line 11 of Example 3.1, also in the previous section, used concatenation to put a colon immediately after the verb. As with lists, the order of concatenation matters: "abc" + "def" is different from "def" + "abc".

String Accumulators

Both the complement() and random_dna() functions of Example 3.2 use string accumulation loops to build their return values. These follow the same pattern as numeric accumulators (Section 2.3) and list accumulators (Section 2.9):

```
accumulator = ""
loop:
    accumulator += string to append
```

⟶ Note: The **empty string** is written as two quotes with nothing in between. Accumulators may start with something else, but as with lists, an empty string is the most common.

⟶ Note: As with lists (see page 84), string accumulators can prepend by writing out an accumulation on the left instead of using the shorthand.

Loop over Characters in a String

Like lists and ranges, strings are a sequence type, so they support **for** loops by looping over each character. The syntax is the same:

```
for character in string:
    body
```

The *character* variable takes on the value of each character in *string* inside the body of the loop. Line 20 of Example 3.2 uses this type of loop to work with each base in the strand of DNA.

Random Strings

To generate random strings of DNA, the `choice()` function of the `random` module is used on line 28 of Example 3.2 to choose a random character from the string `"ACGT"`. The DNA strand is then built with an accumulator. Both `choice()` and `sample()`, listed in Table 3.2, can be used with any sequence type to produce random values.

TABLE 3.2 `random` module: sequence functions

`choice(s)`
Return random element from sequence *s*.
`sample(s, k)`
Return list of *k* random distinct elements from sequence *s*.

⟶ Note: The `sample()` function always returns a list. If *s* contains repeated elements, they may be chosen as "distinct" items of *s* in the sample.

Default Parameter Values

Line 24 of Example 3.2 uses a **default parameter value** of 30 if the function call does not provide an argument. The default value is specified by an assignment after the parameter name:

```
def name(...parameter=default):
```

Function calls that do include the argument will simply ignore the default value.

If a parameter *p* has a default value, all parameters that come after *p* must also specify a default. Otherwise, the interpreter cannot determine which defaults should be used.

EXERCISES

1. Determine the complement of `ATCAGT`.

2. Determine the reverse complement of `TAAAGG`.

3. Determine the output of this code:

```
result = ""
for i in range(1, 5):
    result += str(i**2)
    print(i, result)
```

4. Determine the output of this code:

```
result = ""
for c in "apple":
    print(c, result)
    result = c + result
```

5. Write a `reverse(s)` function to return a copy of the string s in reverse order using a slice. Include a program to test your function.

6. Write a `reverse(s)` function to return a copy of the string s in reverse order using an accumulator rather than a slice. Include a program to test your function.

7. Write a `reverse_complement(dna)` function to return the reverse complement of the given dna. Use the `complement()` function from Example 3.2, and modify `main()` to test your function.

8. Write a `random_rna(length)` function to return a random fragment of mRNA of the given length. Messenger RNA, or **mRNA** is made of the same bases as DNA, except that uracil (U) replaces thymine (T). Include a default length and a program to test your function.

9. Use a string loop to write a function `is_dna(s)` to return `True` if the string s consists entirely of DNA bases and otherwise return `False`. Test your function on random strings of DNA, as well as other, non-DNA strings.

10. Write a function `is_rna(s)` to return `True` if the string s consists only of mRNA bases and otherwise return `False`. Test your function on random strings of mRNA and DNA.

11. A fragment of DNA is a **palindrome** if it is the same as its reverse complement. (This is not the same as a word palindrome.) Write a function `is_palindrome(dna)` to return `True` if the given dna is palindromic, and otherwise return `False`. Use your function to write a program that finds a palindrome of length 10 by testing randomly generated strings of DNA until it finds one.

12. DNA **transcription** can be viewed as producing a strand of mRNA that is the same as the non-template (coding or sense) strand except every thymine (T) is replaced by uracil (U). Write a function `transcription(sense_dna)` to return the corresponding mRNA for *sense_dna*. Test your function on random strings of DNA.

⟶ Note: Transcription actually works on the template (antisense) strand.

13. Write a `vowels(s)` function to return a string consisting only of the vowels in the string s, in the order they appear. Include a program to test your function.

14. Write a `hide_vowels(s)` function to return a copy of the string s with each vowel replaced by a hyphen "-." Include a program to test your function.

15. Write a `random_bits(length)` function to return a string of random bits (0's and 1's) of the given *length*. Include a default value for *length* and a program to test your function.

16. Write a `random_digits(length)` function to return a string of random digits (0 through 9) of the given *length*. Include a default value for *length* and a program to test your function.

17. Write a `random_hex(length)` function to return a string of random hexadecimal digits (0–9, A–F) of the given *length*. Include a default value for *length* and a program to test your function.

18. Write a recursive `reverse(s)` function to return a copy of the string s in reverse order. Include a program to test your function.

19. Write a `romanchar_to_int(c)` function to return the integer value of any single-character Roman numeral c: I, V, X, L, C, D, or M. Return 0 for any other character. Include a program to test your function.

20. (Requires Exercise 19.) Write a `romanpair_to_int(pair)` function to return the integer value of any valid subtractive pair of Roman numerals, such as *pair* = `"IX"`. Include a program to test your function.

21. (Requires Exercises 19 and 20.) Write a `roman_to_int(roman)` function to return the integer value of the Roman numeral given in the string *roman*. Include a program to test your function.

22. (Requires Exercise 21.) Write a `largest_roman(n)` function to return the highest-valued single Roman numeral or subtractive pair with value $\leq n$. Include a program to test your function.

23. (Requires Exercises 21 and 22.) Write an `int_to_roman(n)` function to return a string representing the Roman numeral of n, for any $n > 0$. Include a program to test your function on random integers.

24. (Requires Exercise 23.) Write a `random_roman()` function to return a random valid Roman numeral. Include a program to test your function.

25. The last digit in a 13-digit ISBN book identification number is a **check digit** to help catch transmission errors. Research how to compute the check digit, and then write a `check13(code)` function to return the check digit for the given 12-digit string *code*. Include a program to test your function.

3.3   TEXT IN MEMORY

Section 2.5 described roughly how both integers and floating-point numbers are stored in binary. Binary numbers are used because computer memory is essentially a sequence of on/off switches, and those can be thought of as 0's and 1's.

This raises the question: how is text stored in memory? The basic idea uses two steps: first, assign each character an integer code number. Then, use an encoding to store the integer in binary. Understanding this process allows us to write our own functions to convert integers to strings and vice versa, such as in Example 3.3.

```
1  def digit_to_int(digit):
2      """Return int corresponding to string digit."""
3      return ord(digit) - 48
4
5  def decimal_to_int(digits):
6      """Return int corresponding to decimal string."""
7      result = 0
8      for digit in digits:
9          result = 10 * result + digit_to_int(digit)
10     return result
```

Example 3.3   Convert string to integer.

Unicode

Python strings are sequences of Unicode **code points**, which are integer codes for individual characters. Table 3.3 lists two built-in functions, **chr()** and **ord()** that convert back and forth between characters and integer code points.

TABLE 3.3   Unicode functions

chr(*n*)
Character with Unicode code point *n*.
ord(*c*)
Unicode code point of single character *c*.

Example: **chr**(100) is the letter "d," and **ord**("d") is 100.

Example: Trace the operation of **decimal_to_int**("427").

> **Tracing code** means to simulate the operation of the interpreter on paper. It can help build understanding of an unfamiliar function. Table 3.4 shows one way to trace Example 3.3.

TABLE 3.4   Tracing a function call

result	digit	digit_to_int(digit)	10 * result
0	"4"	4	0
4	"2"	2	40
42	"7"	7	420
427			

Encodings

Unicode is designed to support characters from any written language, so there are more than one million different code points. How code points are stored in binary is determined by a separate **encoding**, such as ASCII or UTF-8.

ASCII (pronounced "ask-ee") is a 7-bit encoding first published in 1963 that became the standard encoding for the English alphabet. Many different variations of "extended ASCII" use 8 bits to match the byte size of hardware.

UTF-8 is a variable-length encoding which extends ASCII, fully supports Unicode, and has become the dominant encoding on the web.

Unicode numeric code points for the English keyboard also match the decimal values of ASCII encodings. Thus, for our purposes, it is safe to imagine text being stored in binary using 8-bit ASCII values with a 0 as the first bit.

Example: ASCII 01010001 encodes "Q" and **ord("Q")** is 81.

Binary 01010001 is equivalent to decimal 81, which is the Unicode code point for "Q."

A useful feature of ASCII is that digits and letters have consecutive codes: digits 0–9 occupy codes 48–57 (decimal), uppercase A–Z have codes 65–90, and lowercase a–z are 97–122.

Escape Sequences

Escape sequences are used to put special characters into strings that cannot be directly typed. All escape sequences begin with a **backslash** "\". Table 3.5 lists some common escape sequences.

Example: \u0301 puts an acute accent on the preceding character.

This example of a **combining character** in Unicode was used on page 96 to create an "á."

Newlines (\n) and tabs (\t), as well as spaces (**chr(32)**), are considered **whitespace** characters, because they result in empty space on the page.

TABLE 3.5 Escape sequences

\n	
Newline	
\t	
Tab	
\"	
Double quote	
\'	
Single quote	
\\	
Backslash	
\u*xxxx*	
16-bit Unicode code point *xxxx* (hex)	
\U*xxxxxxxx*	
32-bit Unicode code point *xxxxxxxx* (hex)	

EXERCISES

1. Determine the value of these expressions:

   (a) `chr(ord("e"))`   (c) `chr(ord("L") + 1)`

   (b) `ord(chr(44))`   (d) `chr(ord("x") + 2)`

2. Determine the value of these expressions:

   (a) `chr(ord("K"))`   (c) `chr(ord("y") - 5)`

   (b) `ord(chr(110))`   (d) `chr(ord("4") + 3)`

3. Trace the operation of `decimal_to_int("1697")`, following Table 3.4.

4. Trace the operation of `decimal_to_int("5623")`, following Table 3.4.

5. Give the range of codes in 7-bit ASCII.

6. Give the range of codes in 8-bit extended versions of ASCII.

7. Look at binary representations of the ASCII codes for corresponding pairs of upper- and lowercase letters (for example, "E" and "e"). Find the pattern, and then use it to explain why the lowercase group does not immediately follow the uppercase group.

8. Explain why curly brackets {} occur at the end of the list of symbols in the index of this text. Where would they be listed relative to alphabetic characters (a–z and A–Z) if the symbols were not listed separately?

9. Incorporate `decimal_to_int()` from Example 3.3 into a complete program.

10. Write a program to display an ASCII table for the characters with codes between 32 and 126.

11. Write a `to_upper(c)` function using `ord()` and `chr()` to return the uppercase letter corresponding to c if c is lowercase, and otherwise returns c without any change. Include a program to test your function.

12. Write a `to_lower(c)` function using `ord()` and `chr()` to return the lowercase letter corresponding to c if c is uppercase, and otherwise returns c without any change. Include a program to test your function.

13. (Requires Exercise 11.) Write an `upper(s)` function to return a copy of the string s with all of its lowercase letters changed to uppercase and the rest unchanged. Use the `to_upper()` function above and include a program to test your function.

14. (Requires Exercise 12.) Write an `lower(s)` function to return a copy of the string s with all of its uppercase letters changed to lowercase and the rest unchanged. Use the `to_lower()` function above and include a program to test your function.

15. Write a `binary_to_int(bits)` function to return the decimal integer value of the string of binary bits. Include a program to test your function.

16. Write a `hexdigit_to_int(digit)` function to return the decimal value of a single valid hex digit. Use this function to write a `hex_to_int(s)` function to return the decimal integer value of the string of hex digits s. Include a program to test your functions.

17. Write an `int_to_digit(n)` function to return the single character representing $n = 0, 1, 2, \ldots, 9$ without using the `str()` function. Include a program to test your function.

18. Write an `int_to_hexdigit(n)` function to return the single hexadecimal character representing n for $n < 16$ without using the `str()` or `hex()` functions. Include a program to test your function.

19. (Requires Exercise 17.) Write an `int_to_decimal(n)` function to return the decimal string corresponding to n for $n > 0$. Use the `int_to_digit()` function above and a string accumulator like this:

    repeat while n is positive:
      concatenate the digit n % 10 on the left
      integer−divide n by 10

    Include a program to test your function.

20. (Requires Exercise 17.) Write an `int_to_binary(n)` function to return the binary string corresponding to n for $n > 0$. Do not use the built-in `bin()` function; instead, use the `int_to_digit()` function above and a string accumulator as described in the previous exercise. Include a program to test your function.

21. (Requires Exercise 18.) Write an `int_to_hex(n)` function to return the hexadecimal string corresponding to n for $n > 0$. Do not use the built-in `hex()` function; instead, use the `int_to_hexdigit()` function above and a string accumulator as described in Exercise 19. Include a program to test your function.

22. Write a function `ascii(msg)` to return a string of the ASCII codes for each character in the string *msg*. For example, `ascii("ABC")` should return the string `"65 66 67"`.

23. Write a `value(word)` function to return the value of *word* based on assigning a=1, b=2, ..., z=26, and computing the sum of the values of the letters in *word*. For example, `value("cat")` is $3 + 1 + 20 = 24$. Assume the word is all lowercase. Include a program to test your function.

24. A **Caesar cipher** with shift n alters every letter in the message by moving forward n steps. For example, if $n = 3$, then $A \rightarrow D$, $B \rightarrow E$, and so on, wrapping around so that $X \rightarrow A$, $Y \rightarrow B$, and $Z \rightarrow C$. Write an `encode(msg, n)` function to encode *msg* using a Caesar cipher with shift n. Assume *msg* consists entirely of uppercase letters, with no spaces or punctuation. Use an `encode_letter(c, n)` function to encode a single letter, and also write a `decode(msg, n)` function to decode *msg*. Include a program to test your functions.

3.4   STRING PROCESSING

Beyond indexing, slicing, and working directly with character codes, Python offers a rich set of methods for processing text. Example 3.4 uses one of these methods to remove punctuation from a string, which is an important step toward isolating the words in a text file. Many common tasks such as spell-checking or word frequency counts rely on having a list of words without whitespace or punctuation.

$\longrightarrow$ Note: `punctuation` from the `string` module is a string containing all ASCII punctuation characters.

Objects and Object Methods

There is new syntax in Example 3.4 that requires explanation. Every data type in Python is an *object* data type, and in order to understand some of

```
1  from string import punctuation
2
3  def remove_punctuation(s):
4      """Return copy of s with punctuation removed."""
5      for c in punctuation:
6          if c in s:
7              s = s.replace(c, "")
8      return s
```

Example 3.4 Remove punctuation.

the functionality of Python strings (and other types), we need to begin to use object-oriented terminology. Chapter 5 will then explore these concepts in greater depth.

A **class** defines the data storage and functionality of an object type. For example, the **str** class defines the string type in Python. An **object** is then a particular **instance** of the class, such as a particular string, that you can send messages to. These messages, known in Python as **method calls**, ask the object to do something. They are like function calls, except that they are called on an object and have access to the data of that object.

Programming that focuses on using objects is called **object-oriented**. The programming you have done to this point without thinking in terms of objects is usually called **imperative** or **procedural** programming. By design, Python supports multiple paradigms, including both imperative and object-oriented programming.

Syntax: Method Calls

The syntax to call an object's method is known as **dot notation**:

```
object.method(argument1, argument2, ...)
```

This code will cause *object* to execute *method* with the given arguments. Compare this with the syntax of a function call in Section 1.2: for a method call, you must ask a particular object to perform the method, and there is a dot (period) between the object and the name of the method. Calling the same method from different objects will usually produce different results.

For example, s.replace(c, "") on line 7 of Example 3.4 calls the replace() method on the string s with arguments c and the empty string.

String Methods

Python provides a very rich **str** class with a wide assortment of methods. Tables 3.6, 3.7, and 3.8 list a few of them, separated into three different categories. Notice that none of the string manipulation methods in Table 3.6 change the string they are called on; they all return a modified copy of the string.

TABLE 3.6   String manipulation methods

`s.upper()`
Return uppercase copy of s.
`s.lower()`
Return lowercase copy of s.
`s.capitalize()`
Return copy of s with first letter uppercase, the rest lowercase.
`s.title()`
Return copy of s with each word capitalized.
`s.replace(old, new)`
Return copy of s with each occurrence of *old* replaced by *new*.
`s.strip(chars)`
Return copy of s with leading and trailing characters in *chars* removed; strips whitespace if *chars* omitted.

TABLE 3.7   String search methods

`s.startswith(t)`
Return **True** if *t* is a prefix of *s*.
`s.endswith(t)`
Return **True** if *t* is a suffix of *s*.
`s.count(t)`
Return number of occurrences of *t* as a substring of *s*.
`s.find(t, i, j)`
Return first index where *t* is a substring of `s[i:j]` or -1 if not found. Either *j* or both *i* and *j* may be omitted.

Mutable and Immutable Objects

If you search the Python documentation, you will see there are *no* methods that modify the string they are called on. The reason is that strings are immutable.

An object's **state** is the data that it stores. For example, a string stores its length and the character in each position. An object that may change its state is called **mutable**, while objects that can never change their state are

TABLE 3.8   String inspection methods

`s.isalpha()`
Return **True** if all characters in s are alphabetic.

`s.isupper()`
Return **True** if all characters in s are uppercase.

`s.islower()`
Return **True** if all characters in s are lowercase.

`s.isdigit()`
Return **True** if all characters in s are digits.

---

immutable. Strings are immutable, so they can never change the data they store. At this point, we have only seen one mutable type, as summarized in Table 3.9.

TABLE 3.9   Mutable and immutable types

Mutable	Immutable
list	int
	float
	bool
	str

⟶ Note: Because strings are immutable, they do not support the indexed or slice assignments described for lists (which are mutable) on page 80.

Reassignment with Immutable Types

In Example 3.4, we want to remove punctuation, so how is that possible if strings are immutable? The answer is the same as it is for other immutable types such as integers: if you want an integer value such as *count* to change, you *assign* it to a new value:

```
count = count + 1
```

The same is true for strings and any immutable type. If you want to modify a string, **reassign** it to a new value:

```
variable = new value
```

View it as if the variable now names the new value instead of the old, as in Fig. 3.2. The *old value* object doesn't change.

This explains why line 7 of Example 3.4 reassigns s:

```
s = s.replace(c, "")
```

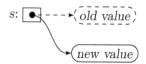

Figure 3.2 Reassignment.

Just calling s.replace() without the assignment would create a new string but not change the value of s.

EXERCISES

1. Let s = "circadian rhythms". Determine the value of these expressions:

   (a) s.upper()

   (b) s.title()

   (c) s.startswith("circa")

   (d) s.endswith("thm")

2. Let t = "Fun with Python". Determine the value of these expressions:

   (a) t == t.capitalize()

   (b) t.title() == t

   (c) t.isalpha()

   (d) t[4:8].islower()

3. Let s = "encore presentation". Determine the value of these expressions:

   (a) s.replace("e", "-")

   (b) s.find("present")

   (c) s.count("n")

   (d) s.count("en")

4. Let t = "open-source software project". Determine the value of these expressions:

   (a) t.find("p")

   (b) t.find("p", 2)

   (c) t.replace("soft", "hard")

   (d) t.count("so")

5. Determine the final value of s after this code runs:

```
s = "string methods with strings"
s.replace("string", "file")
s.upper()
```

6. Determine the final value of s after this code runs:

```
s = "Python is named after Monty?"
i = s.find("Monty")
s = s[i:-1] + " " + s[:6]
```

7. Determine the result of executing this code:

```
s = "currant jam"
s[4] = "e"
```

8. Determine the result of executing this code:

```
s = "ambient music"
s[:7] = "discreet"
```

9. Write one line of code to change the string `film` to capitalize each word in it.

10. Write one line of code to change the string `message` so that it is all uppercase.

11. Write one line of code to change the string `s` so that all hyphens are removed.

12. Write one line of code to change the string `t` so that all spaces are removed.

13. Write one line of code to add an exclamation point to the end of the string `s`.

14. Write one line of code to change the last character of the string `s` to a question mark.

15. Incorporate `remove_punctuation()` from Example 3.4 into a complete program to test it.

16. Write a function `keep_apostrophes(s)` to return a copy of `s` with all punctuation except for apostrophes removed. (This keeps contractions intact.) Include a program to test your function.

17. Write a function `punctuation_only(s)` to return a copy of `s` that retains only punctuation characters. Include a program to test your function.

18. Write a function `alpha_only(s)` to return a copy of `s` that retains only alphabetic characters. Include a program to test your function.

19. Write the function `mycapitalize(s)` to mimic the string `capitalize()` method without calling it. Test your function by comparing it with the method.

20. Write a function `is_dna(s)` to return `True` if the string `s` consists entirely of DNA bases (see Section 3.2) and otherwise return `False`. Allow `s` to be upper, lower, or mixed case. Include a program to test your function.

21. Write a function `is_rna(s)` to return `True` if the string `s` consists only of mRNA bases (see page 104) and otherwise return `False`. Allow `s` to be upper, lower, or mixed case. Include a program to test your function.

22. Write a function `transcription(sense_dna)` to return the corresponding mRNA for *sense_dna* (see page 104). Allow *sense_dna* to be upper, lower, or mixed case. Test your function on random strings of mixed-case DNA.

23. Write a function `is_ly(s)` to return `True` if *s* is a single word (no spaces) ending in "-ly," like many adverbs. Include a program to test your function.

24. Write a function `is_identifier(s)` to return `True` if *s* consists entirely of alphabetic characters, underscores (_), and, except for the first character, the digits 0–9; otherwise, it returns `False`. Include a program to test your function.

25. Write a function `is_palindrome(s)` to return `True` if *s* is a **palindrome**, that is, a phrase that reads the same backward as forward, ignoring any punctuation or whitespace, as well as whether characters are upper- or lowercase. For example, "Madam, I'm Adam!" should return `True`. Include a program to test your function.

26. Write `encode()` and `decode()` functions for the Caesar cipher of Exercise 24 on page 110 that maintain upper- and lowercase letters, as well as spaces and punctuation. Include a program to test your functions.

HOW-TO: OUTPUT FORMATTING

By default, Python displays floats with a variable number of decimal places, and integers are left-aligned rather than right-aligned in tables of values. The string `format()` method, listed in Table 3.10, provides control over issues like these.

TABLE 3.10   String format method

`s.format(args)`
Return copy of *s*, formatting *args* into its replacement fields.

Example: `print("{:3d}    ${:>7.2f}".format(year, balance))`

The arguments to the `format()` method (`year` and `balance`) are substituted into the **replacement fields** delimited by braces `{}` in the string. This example has two replacement fields; all other characters in the string (spaces and the dollar sign) are literal.

`{:3d}` formats `year` as a decimal integer in a minimum of 3 spaces that are right aligned.

`{:>7.2f}` formats `balance` as a fixed-point value with 2 digits after the decimal point and minimum width 7, right aligned by the `>`.

Consult the documentation for a complete list of formatting options.

The `print()` function also has two optional arguments, given in Table 3.11, that provide more flexibility. The default values of **sep** and **end** can be changed by specifying them in the call.

TABLE 3.11   Print options

```
print(objects, sep=" ", end="\n")
```
Print *objects*, separated by **sep** and followed by **end**.

Example: `print(x, y, z, sep="")`

Prints x, y, and z with no space between them (and the default newline at the end).

3.5   LISTS OF STRINGS

In a sense, there is nothing special about lists of strings in Python, as lists may hold objects of any type. In fact, the `conjugate()` function in Example 3.1 returned a list of the conjugated verb forms at the beginning of this chapter. At the same time, many common tasks are based on lists of strings or characters, and Python offers powerful techniques for working with them. The `wordlist()` function in Example 3.5 performs a fundamental task: it returns the list of words from any string of text, after first removing punctuation.

⟶ Note: To run Example 3.5, include the `remove_punctuation()` function from Example 3.4.

```python
1  def wordlist(text):
2      """Return list of words in string."""
3      return remove_punctuation(text).split()
4
5  def main():
6      examples = ["Is this a good example?",
7                  "Yes, of course it is!",
8                  "Ok - thanks."]
9      for text in examples:
10         print(wordlist(text))
11
12  main()
```

Example 3.5   List of words in text.

Split and Join

Two of the main tools for working with lists of strings are the `split()` and `join()` string methods listed in Table 3.12. They perform opposite tasks: `split()` breaks one string into a list of strings, whereas `join()` glues a list of strings together into one string.

TABLE 3.12   String split and join

`s.split(separator)`
Return list of words in string *s* split by optional *separator*; splits at whitespace if *separator* omitted.

`separator.join(list_of_strs)`
Return concatenation of strings in *list_of_strs* with *separator* between each string.

⟶ Note: `join()` is called on the separator string, not the list.

⟹ Caution: `split()` works differently than `split(separator)`. See the examples below.

Example: `"Hello, world.".split()` is `["Hello,", "world."]`.

> A `split()` based on whitespace puts punctuation with whatever word it is adjacent to.

Example: `"   First      Last ".split()` is `["First", "Last"]`.

> Splitting on whitespace considers any amount of whitespace as one separator and ignores leading and trailing spaces.

Example: `"123-45---6".split("-")` is `["123", "45", "", "", "6"]`.

> However, when the *separator* parameter is used, consecutive separators result in empty strings in the list, since there is nothing between those separators.

Example: `"".join(["one", "two", "three"])` is `"onetwothree"`.

> Joining on the empty string concatenates the strings in the list.

Example: `" ".join(["one", "two", "three"])` is `"one two three"`.

> Changing the separator to a space puts one space between each of the strings.

Lists of Examples

The main program in Example 3.5 demonstrates a useful technique for creating a list of examples to test the `wordlist()` function. The **for** loop in line 9 is a loop over the list of strings in `examples`. This makes it easier to repeat tests than using input to ask for examples.

List Mutators

In the previous section, we learned about method calls and mutable types. Lists are mutable and have several methods that change their state; such methods are known as **mutators**. Some of the most common list mutators are given in Table 3.13.

TABLE 3.13   List mutators

items.append(*x*) Append *x* to the end of list *items*.
items.insert(*i*, *x*) Insert *x* into list *items* at index *i*, shifting existing elements to the right.
items.pop() Remove and return the last item in *items*.
items.pop(*i*) Remove and return *items*[*i*].
items.remove(*x*) Remove the first occurrence of *x* in *items*. Raises **ValueError** if *x* not in *items*.
items.reverse() Reverse the order of the elements in *items*.
items.sort() Sort the list *items* in increasing order.

⟹ Caution: Most of the methods in Table 3.13 return **None**—the special value returned by functions with no other return value (see page 46 in Section 2.1). In particular, the **reverse()** and **sort()** methods *do not* return the reversed or sorted list. Only **pop()** returns the item it removes.

The **random** module also includes a list mutator, given in Table 3.14. It also returns **None**, because it shuffles the list without creating a new list.

TABLE 3.14   **random** module: list shuffle

shuffle(*items*) Randomly shuffle the elements in *items*.

Mutating Strings as Lists of Characters

Strings are immutable, but converting a string to a mutable list of characters allows those characters to be changed or rearranged. The built-in **list()** type converter returns a list of characters when called on a string.

Example: list("mouse") is ["m", "o", "u", "s", "e"].

Once converted to a list of characters, list operations can be used to change the characters. Then a join() on the empty string will convert the list back to a string.

Example: Change the last character of the string s to an exclamation point.

The following steps illustrate how to use this technique:

```
chars = list(s)
chars[-1] = "!"
s = "".join(chars)
```

Remember that s must be reassigned in order to change. There may also be simpler ways to reassign s to produce this particular result.

Sequence Methods

Table 3.15 lists two other list methods that are not mutators, which are in fact available for any sequence type. As noted in Table 3.7, the count() method counts substrings when called on strings. The index() method is similar, but usually the specialized find() method (also in Table 3.7) is easier to use on strings.

TABLE 3.15 Sequence search methods

items.count(*x*)   Return number of occurrences of *x* in *items*.
items.index(*x*, *i*, *j*)   Return first index where *x* occurs in *items*[*i*:*j*] or raises **ValueError** if not found. Either *j* or both *i* and *j* may be omitted.

EXERCISES

1. Determine the value of these expressions:

    (a) "Chapter 1: Introduction".split()

    (b) "$1.99 per item, today only!".split()

    (c) "".join(["h", "e", "l", "l", "o"])

    (d) " ".join(["central", "processing", "unit"])

2. Determine the value of these expressions:

    (a) "On the road -- again.".split()

    (b) "19.5,3.48,1.997,20.3".split(",")

    (c) "".join(["reading", "and", "writing"])

    (d) "-".join(["299", "01", "4185"])

3. Determine the error(s) in these expressions:

   (a) `split("aardvark ant anteater antelope")`

   (b) `" ".join("glue", "these", "words", "together")`

4. Determine the error(s) in these expressions:

   (a) `["apple", "orange", "pear", "tangerine"].split()`

   (b) `"".join([1, 2, 3, 4])`

5. Let items = `[1, 3, 5, 7, 9, 11]`. Determine the new value of items after each of these method calls; start with the original value of items for each.

   (a) `items.pop()`   (c) `items.remove(3)`

   (b) `items.append(2)`

6. Let colors = `["red", "green", "blue"]`. Determine the new value of colors after each of these method calls; start with the original value of colors for each.

   (a) `colors.reverse()`

   (b) `colors.insert(1, "purple")`

   (c) `colors.sort()`

7. Let word = `"deducing"`. Determine the value of word after this code runs:

   ```
   chars = list(word)
   chars.insert(5, "t")
   word = "".join(chars)
   ```

8. Let word = `"lexical"`. Determine the value of word after this code runs:

   ```
   chars = list(word)
   chars.sort()
   word = "".join(chars)
   ```

9. Write code to change the second character of a string named phrase to a hyphen.

10. Write the pattern for a list accumulation loop using `append()`. Describe what is different from the pattern in Section 2.9.

11. Write a function `count_words(text)` to return the number of words in the string *text*, excluding punctuation. Include a program to test your function.

12. Write a function `first_word(text)` to return the first word in the string *text*, excluding punctuation. Include a program to test your function.

13. Write a function `count_word(word, text)` to return the number of occurrences of *word* in the string *text* as a word, not a substring. Ignore punctuation as well as upper-/lowercase differences when counting, and include a program to test your function.

14. Write a function `longest_word(text)` to return the longest word in the string *text*, excluding punctuation. Include a program to test your function.

15. Write the function `mytitle(s)` to return a copy of *s* with each word capitalized, without using the string `title()` method. Assume *s* is a simple string with one space between each word. Test your function by comparing it with the method.

16. Write a function `acronym(phrase)` to return the acronym of the given *phrase*. For example, the acronym of "as soon as possible" is "ASAP." Include a program to test your function.

17. Write a function `decode_ascii(msg)` to decode the `ascii()` encoding of Exercise 22 on page 110. For example, `decode_ascii("65 66 67")` should return the string `"ABC"`. Include a program to test your function.

18. Write a function `mutate(dna)` to return a copy of the string *dna* with a random mutation of one of its bases to a different base. Include a program to test your function.

19. Write a function `jumble(word)` to return a scrambled version of *word* (as a string) by randomly shuffling its letters. Include a program to test your function.

20. Write a function `jumble_words(text)` to return a scrambled version of the string *text* by randomly shuffling its words after removing punctuation. The function should return a string. Include a program to test your function.

21. Write a function `sortchars(s)` to return a string with the same characters in it as the string *s*, except in alphabetical order. For example, `sortchars("program")` is `"agmoprr"`. Include a program to test your function.

22. Write a function `is_anagram(word1, word2)` to return True if *word1* is an anagram of *word2*; otherwise, it returns False. Two words are **anagrams** of each other if they consist of precisely the same letters. For example, "pass" and "spas" are anagrams. Include a program to test your function.

23. Write an `is_phrase_anagram(phrase1, phrase2)` function to return True if *phrase1* is an anagram of *phrase2*, ignoring whitespace, punctuation, and upper-/lowercase; otherwise, it returns False. Include a program to test your function.

3.6   READING TEXT FILES

Being able to access text from a file opens up a whole new range of possibilities for interesting programs. Example 3.6 does a spelling check, testing to see if the input word appears in a list of words from a dictionary. It contains another important basic function: `contents()` returns the entire contents of a text file as one Python string.

Example 3.6 assumes the dictionary file contains no punctuation and all words in it are lowercase. A sample dictionary file is available at the book website `http://www.central.edu/go/conciseintro/`.

—→ Note: To run Example 3.6, put the dictionary file in the same folder as the Python program.

```
1  # speller.py
2
3  def contents(filename):
4      """Return contents of text file as string."""
5      with open(filename) as f:
6          return f.read()
7
8  def main():
9      validwords = contents("dictionary.txt").split()
10     word = input("Enter a word: ")
11     if word.lower() in validwords:
12         print("Looks ok")
13     else:
14         print("Not in this dictionary")
15
16 main()
```

Example 3.6   Spell checker.

Files

Python makes it particularly easy to access files. A **file** is a collection of data stored by the operating system in such a way that it can be retrieved by its name and location in a directory or folder hierarchy. Files are an abstraction of the operating system, meaning the operating system allows us to think of files as identifiable things without having to worry about the details of how they are actually stored on disk. **Reading** a file means to look at the data in the file without making any modifications, which is enough access to perform many useful tasks.

Python File Objects

Files are accessed in Python through **file objects**. The built-in **open()** function, listed in Table 3.16, is used to get file objects for access to existing files. It has many options, some of which will be described later, but to open a text file for reading, all you need to do is provide the file name (or an absolute or relative path) as a string parameter.

TABLE 3.16 Open text file

open(*pathname*)
Return file object for reading text file at *pathname*. May raise **OSError**.

Syntax: **with** Statements

Because things can go wrong when opening a file—for example, it may not exist or it may be corrupted—it is important to open files carefully so that errors may be handled gracefully. The recommended way to open a file in Python is to use a **with** statement:

```
with expression as variable:
    body
```

What happens when this statement executes is complicated, but when it is used to open a file, the steps are approximately like this:

1. An attempt is made to open the file.

2. If successful, the open file object is assigned to *variable*.

3. The *body* is executed.

4. The file is closed at the end, even if something goes wrong in the *body*.

Because the file is closed at the end, the file object may only be used inside the body of the **with**. In general, limit computations done while the file is open to only what is necessary. In the case of Example 3.6, we just need a list of words from the dictionary, so the file is closed as soon as that list is returned.

Methods of File Objects

Given a file object *f* for an open file, the methods in Table 3.17 may be used to read the contents. The **read()** method is usually enough, unless the file is too large to fit in memory or you specifically need to work with its individual lines. Any of them may raise an **OSError**.

TABLE 3.17   Text file read methods

`f.read()`
Return string containing entire contents of file referenced by f.

`f.readlines()`
Return list of all lines from file referenced by f. Each line includes its trailing newline, \n, if present.

`f.readline()`
Return next line from file referenced by f, including the trailing newline, \n. At end of file (only), return empty string.

File Loops

To loop over all the lines in a file, the file object f can be put in a **for** loop:

```
for line in f:
    body
```

This is equivalent to looping over `f.readlines()` but is simpler to read. Because it is like `readlines()`, each line contains its trailing newline, \n.

Combining Function and Method Calls

A function (or method) should perform one well-defined task and possibly return one value. Part of the art of programming is learning to combine existing function and methods calls in new ways to solve new problems. Line 9 of Example 3.6 calls the `split()` method on the string returned by the new `contents()` function:

$$\textbf{return } \underbrace{\texttt{contents("dictionary.txt")}}_{\text{string}}\texttt{.split()}$$

This is the same idea as a nested function call, introduced in Section 1.8 on page 36, but it looks different because of the method call syntax. Similarly, methods may be called one after another, as long as each return type fits the next call. Watching return types is the key to both reading and writing expressions like these in Python.

Example: `f.read().split()[0]`

Assuming f is an open file object, the types are:

$$\underbrace{\underbrace{\texttt{f.read()}}_{\text{string}}\texttt{.split()}\,\texttt{[0]}}_{\text{list}}$$

so the expression gives the first word read from the file.

Downloading Text Files

Techniques from this section allow you to read and interact with the content of any text file. Project Gutenberg (http://www.gutenberg.org/) provides access to a wide range of books in the public domain. Choose "Plain Text UTF-8" format.

If you have difficulty opening a file in UTF-8 format, specify the encoding in the call to **open()**:

```
open(filename, encoding="utf-8")
```

EXERCISES

1. Determine the values of these expressions:

    (a) "AND, OR, and NOR gates".lower().split()

    (b) "3x + 12 = 4z".replace("x", "y").upper()

2. Determine the values of these expressions:

    (a) " ".join(["graphics", "processing", "unit"]).title()

    (b) "Arithmetic Logic Unit".split().pop(1).upper()

3. Determine the error(s) in this expression:

    "Red Green Blue".split().lower()

4. Determine the error(s) in this expression:

    ["AND", "OR", "NOT", "XOR", "NAND"].sort().pop()

5. Write a function **linelist(*filename*)** to return a list of the lines in the given file. Include a program to test your function.

6. Write a function **count_lines(*filename*)** to return the number of lines in the given file. Include a program to test your function.

7. Write a function **count_words_in_file(*filename*)** to return the number of words in the given file after punctuation is removed. Include a program to test your function.

8. Write a function **random_word_in_file(*filename*)** to return a random word from the given file after punctuation is removed. Include a program to test your function.

9. Write a function **longest_word_in_file(*filename*)** to return the longest word in the given file after punctuation is removed. Include a program to test your function.

10. Write a function `avg_word_length(filename)` to return the average length of the words in the given file after punctuation is removed. Include a program to test your function.

11. Write a program to find the word with the highest value in a given file, as defined in Exercise 23 on page 110. Ignore punctuation, and check all words as if they were all lowercase. Use appropriate functions.

12. Write a program to find the longest word that is a palindrome in a given file, ignoring upper-/lowercase and punctuation. (See Exercise 25 on page 116, but do not ignore whitespace for this problem.) Use appropriate functions.

13. Write a program that chooses a secret random word from a dictionary, prints a jumbled version of the word (see Exercise 19 on page 122), and then asks the user to guess the word. Count the number of guesses used.

14. (Requires Exercise 13.) Long words chosen randomly in the previous problem are hard to guess. Write a function `shortwords(limit, words)` to return a list of the words in the list *words* whose length does not exceed *limit*. Use this function to write an easier version of the guessing program.

15. Write a function `anagrams(word)` to return a list of all the words in a dictionary file that are anagrams of *word* and different from it. (See Exercise 22 on page 122.) Include a program to test your function.

16. Write a function `spellcheck(filename)` to return a list of all the words in the given file that do not appear in the dictionary file. Include a program to test your function. Caution: this program may cause IDLE to hang with large files.

17. (Requires Exercise 16.) Explore ways of improving the function from the previous exercise with either binary search (see Section 2.10) or Python sets (see the documentation).

PROJECT: WORD-GUESSING GAME

These exercises lead up to developing a game in which the object is to guess a word by guessing the letters that appear in it. Letters guessed correctly are shown where they appear in the word. For example, output in the middle of a game might be:

```
The word: -est--e
You have guessed: enstr
You have 4 guesses remaining.
Guess a letter:
```

The following exercises outline one way to adopt a spiral approach (see page 71) for developing this program.

EXERCISES

1. Write a program to choose a random word from a dictionary file and allow the user to guess a letter in the word six times. Inform the user of the number of guesses remaining. At the end, display the secret word.

2. Keep track of the letters guessed so far, and inform the user of their guesses before each new guess.

3. Inform the user whether or not the letter that was guessed is in the word, and only reduce the number of guesses remaining in the case of an incorrect guess.

4. Write a `template(secret, guesses)` function to return what is known about the *secret* based on the letters in *guesses*. In the example above, `template("festive", "enstr")` returns `"-est--e"`. Report the template to the user before each guess, and stop the program when the template becomes the same as the secret.

5. Display the letters guessed so far in alphabetical order.

6. Do not penalize the user for repeating a guess. If the user guesses a letter that has already been guessed, print a reminder and allow another guess.

PROJECT: FLASH CARDS

The following exercises outline development of a flashcard quiz program. Begin by creating a text file containing pairs of items that might be on the front and back of a set of flashcards. Use a character such as a colon as a delimiter to separate the front from the back of each card. For example, Latin vocabulary cards might look like this:

```
life:vita, -ae
gate:porta, -ae
always:semper
...
```

Save the file with a name like `cards.txt` in the same folder in which you will write your program.

When using a literal value such as `":"` in a specific way like this, use a **named constant** so that its meaning is apparent:

```
DELIMITER = ":"
```

Constants used by only one function can be put at the beginning of the function; otherwise, include them at the top of the file with **import** statements.

The steps below represent each flashcard as a two-element list. For example, the first card above will become ["life", "vita, -ae"] in the program. The front of the card will always be the first item in the pair, and the back of the card second.

EXERCISES

1. Write a function pairs(*text*) to return a list of front-back pairs for all of the cards in the string *text*. Assume there is one card per line in *text*, and that each line contains two parts separated by the delimiter. Using the examples above, the list returned should begin:

   [["life", "vita, -ae"], ["gate", "porta, -ae"], ...]

   Include a main() to read cards.txt and test your function.

2. Write a function quiz(*cards*) to quiz the user over each of the card pairs in the list *cards*. Shuffle the cards, and then for each card, show the front and ask the user for the back. Indicate whether or not each response is correct, and show the user the correct response when they are incorrect.

3. Write a function get_responses(*correct*, *limit*) to allow the user to answer up to *limit* number of times until providing the *correct* response for a single card. (You do not need to repeat the question.) Modify your program to use this function.

4. Modify your program to keep track of the number of cards for which the user eventually provided a correct response. Report overall performance at the end.

HOW-TO: READING CSV FILES

Many public data sets, such as those at data.gov, are available in **CSV** format, which stands for comma-separated values. CSV files are like spreadsheets, where each line is a row, and column entries are separated by commas. For example, a demographic data file might contain entries like this, with one row of data for each zip code:

```
90001,57110,26.6,28468,28642,12971,4.4
90002,51223,25.5,24876,26347,11731,4.36
90003,66266,26.3,32631,33635,15642,4.22
90004,62180,34.8,31302,30878,22547,2.73
90005,37681,33.9,19299,18382,15044,2.5
```

The csv module provides a convenient reader() function, listed in Table 3.18, that allows Python programs to work with one row of a CSV file at a time. It can be used in a loop to process each row, as shown in the following fragment.

```
from csv import reader

def readcsv(filename):
    with open(filename, newline="") as f:
        for row in reader(f):
            process(row)

def process(row):
    # work with one row of data
```

The row object obtained from reader() is a Python list containing each element in that row. In most cases, it will be a list of strings, so you may need to convert some entries to a numeric type.

⟶ Note: Downloaded files may need to be opened with UTF-8 encoding; see page 126.

TABLE 3.18 csv module: reader

reader(f)
Return one line at a time from open CSV file object f as a Python list.

Consult the csv module documentation for more information.

3.7 HANDLING EXCEPTIONS

No one likes it when programs crash. And you have probably been working hard to fix errors that have caused programs to crash. But some errors are not the result of incorrect programming, such as trying to open a file that was accidentally deleted. In cases like this, it would be better to inform the user that the file cannot be opened rather than have the program crash.

Example 3.7 updates the program from the last section to handle situations like this. It illustrates several important points about handling exceptions.

Exceptions

An **exception** is an error detected during program execution, as opposed to a syntax error, which can be detected by the interpreter prior to execution. An exception is **raised** when the error is encountered, and at that point it may or may not be **handled** by the running program. Any function in the chain of calls leading to the exception is eligible to handle it, beginning with the function in which the error occurred. Exceptions that are not handled by any function cause the program to terminate with an error message.

Exceptions are handled by putting the code that might raise them in a **try** block.

```
1  # speller2.py
2
3  def contents(filename):
4      """Return contents of text file as string.
5
6      Args:
7          filename - name of file
8      Returns:
9          contents of file as one string
10     Raises:
11         OSError if error opening or reading file
12     """
13     with open(filename) as f:
14         return f.read()
15
16 def main():
17     try:
18         validwords = contents("dictioary.txt").split()
19     except OSError as err:
20         print(err)
21         print("Stopping, unable to read dictionary file.")
22         return
23     word = input("Enter a word: ")
24     if word.lower() in validwords:
25         print("Looks ok")
26     else:
27         print("Not in this dictionary")
28
29 main()
```

Example 3.7 Spell checker with error handling.

Syntax: **try-except**

The syntax of a basic **try** statement is:

```
try:
    body
except ExceptionType as errorname:
    errorbody
```

In a **try** block, the *body* is executed, and there are three possibilities:

If there are no errors, then execution continues normally, skipping the **except** clause and *errorbody*.

If an *ExceptionType* error occurs, then its message is assigned to *errorname*, and *errorbody* is executed. Normal execution continues after *errorbody*. Some common error types are listed in Table 3.19.

If any other error occurs, then the error is given to the caller to handle.

The "**as** *errorname*" is optional.

More complex **try** statements are possible, including multiple **except** clauses, as well as optional **else** and **finally** clauses. Consult the Python documentation for more information.

TABLE 3.19   Common exception types

IndexError
A sequence index is outside the valid range.
KeyError
A key does not have an entry in a dictionary.
OSError
Error raised by operating system attempting to access a file.
TypeError
The operation is not supported for an object of this type.
ValueError
An object has the correct type but an inappropriate value.

Catch Exceptions Where You Know What to Do

The spelling checker gets a list of valid words from a dictionary file, so there is a possibility something could go wrong while either opening or reading the file. If either the **open()** or **read()** function raises an exception, the chain of

$$\texttt{main()} \longrightarrow \texttt{contents()} \longrightarrow \begin{array}{l} \texttt{open()} \\ \text{or } \texttt{f.read()} \end{array}$$

Figure 3.3 Chain of function calls.

function calls will be as in Fig. 3.3, meaning that `contents()` has the first chance to handle the exception, and if it does not, then `main()` can.

Example 3.7 handles the exception in `main()`, in lines 17–22. The reason is that if a spell-checking program cannot read the dictionary, then there is no point in continuing the program. In other words, `main()` is in a good position to know what to do if there is an error. With this design, there is no change to the `contents()` function from Example 3.6, except for its docstring (discussed shortly).

The alternative would be to move the handling code to the `contents()` function. The question is, does the `contents()` function know what to do if the file cannot be opened or read? A reasonable answer is for it to return an empty string—if the file cannot be read, then the contents are empty. Exercise 2 explores this alternative.

This situation is not unusual: there is often more than one way to handle exceptions in a program design. Either way, cooperating functions need to know what to expect from each other, which is why docstrings often contain more information than a one-line summary of what the function does.

Function Docstrings

The `contents()` function in Example 3.7 illustrates a more complete docstring than previous examples. After a one-line summary followed by a blank line, function docstrings should describe their arguments, return value, and any exceptions that are raised. This way, functions that call `contents()` can be aware that it raises `OSError`, indirectly, because it calls `open()` and `f.read()`.

Docstrings may be printed in the interpreter with the built-in `help()` function.

Example: `help(contents)` displays the docstring of the `contents()` function in Example 3.7, as long as the interpreter has run this program since its last restart.

Syntax: Raising an Exception

An exception can be raised at any time with the **raise** statement:

```
raise exception
```

If the *exception* is omitted, the most recent exception in the current context is raised again. This is useful in **except** clauses when an exception handler performs some cleanup but also wants to alert its caller. More complex forms of the **raise** statement are described in the documentation.

Prevention First

Finally, it is still good practice to prevent exceptions whenever you can, rather than relying heavily on exception handlers. We have seen functions that raise **ValueError**, **KeyError**, and **IndexError** before, but because these errors are largely preventable (for example, don't use an invalid index in a list), they have not required exception handling. Use **try** blocks only when necessary.

EXERCISES

1. Modify Example 3.7 to inform the user and stop if the list of valid words from the dictionary is empty, since in that case, there will be no way to look up the word that is entered.

2. Rewrite Example 3.7 so that **contents()** returns an empty string instead of ignoring exceptions (see page 133). Discuss the tradeoffs between this and the original design.

3. Rewrite Exercise 5 from Section 3.6 to handle exceptions. Include a full docstring for the **linelist()** function.

4. Rewrite Exercise 8 from Section 3.6 to handle exceptions. Include a full docstring for the **random_word_in_file()** function.

5. Rewrite Exercise 9 from Section 3.6 to handle exceptions. Include a full docstring for the **longest_word_in_file()** function.

6. Rewrite Exercise 10 from Section 3.6 to handle exceptions. Include a full docstring for the **avg_word_length()** function.

7. Rewrite Exercise 11 from Section 3.6 to handle exceptions. Include a full docstring for at least one function.

8. Rewrite Exercise 12 from Section 3.6 to handle exceptions. Include a full docstring for at least one function.

9. Rewrite Exercise 15 from Section 3.6 to handle exceptions. Include a full docstring for the **anagrams()** function.

10. Rewrite Exercise 16 from Section 3.6 to handle exceptions. Include a full docstring for the **spellcheck()** function.

HOW-TO: WRITING TEXT FILES

Writing to text files is initiated by opening the file with a different mode. The default mode is `"r"` for reading; several options for writing are listed in Table 3.20. Consult the documentation for a complete list.

TABLE 3.20   Text file modes

`open(pathname, mode="w")`
Return file object for writing to text file at *pathname*. May be used to create a new file, or if the file exists, its contents are deleted first.
`open(pathname, mode="x")`
Return file object for exclusively creating and writing to text file at *pathname*. Fails to open if file already exists.
`open(pathname, mode="a")`
Return file object for appending to text file at *pathname*.
`open(pathname, mode="r+")`
Return file object for reading and writing to text file at *pathname*.

Once a file has been opened with a mode that enables writing, the `write()` method, listed in Table 3.21, may be called on the file object.

TABLE 3.21   Text file write method

`f.write(s)`
Write string *s* to file referenced by *f*. Return number of characters written. May raise `OSError`.

3.8   DICTIONARIES

Imagine writing a program to count the number of times each word appears in a large file. We could store the counts in a list, but then how do we keep track of which word uses which index? We could store the words in a separate second list, so that the count for the word in `words[i]` is in `counts[i]`, but then we would have to search the entire word list every time we came to another word to count.

A Python dictionary is the right tool for this task because it can store a separate integer count for each word being counted. Example 3.8 shows how to use a dictionary to obtain a frequency count for any list of items.

Dictionaries Store Key-Value Pairs

A **dictionary** is a mapping that stores keys with associated values. For every **key** in a dictionary, there is exactly one **value** associated with it. This

```
1  def frequency(items):
2      """Return count of each item in list of items."""
3      count = {}
4      for item in items:
5          if item in count:
6              count[item] += 1
7          else:
8              count[item] = 1
9      return count
```

Example 3.8 Word frequency.

relationship is directional, meaning that from a key, you can look up a value:

$$key \longrightarrow value$$

but not the reverse. Because they associate keys with values, dictionaries are also known as **associative maps**.

Dictionaries are written inside braces {} (also known as curly brackets), with a colon between each key-value pair. For example, in this dictionary:

```
age = {"Alice":35, "Bob":39, "Chuck":35, "Dave":34}
```

the keys are the names, and the ages are the values. Notice that each name has exactly one age, but the same age might be associated to more than one name. A pair of empty brackets {} denotes an **empty dictionary**.

$\longrightarrow$ Note: Dictionary key-value pairs are not stored in any particular order, so dictionaries are *not* a sequence type like strings or lists.

Example: Population of cities (in millions)

```
population = {"Chicago":2.7, "New York":8.17, "Rome":2.87,
              "Paris":2.24, "London":8.78}
```

Example: HTTP response status codes

```
status = {200:"ok", 404:"not found", 400:"bad request"}
```

Dictionary Operations

Python dictionaries use index-like notation to refer to the value associated with *key* in a dictionary *d*:

```
d[key]
```

TABLE 3.22 Dictionary operations

`d[key] = value`
Set value for *key* in dictionary *d* to be *value*.

`d[key]`
Get value for *key* in dictionary *d*. Raises `KeyError` if *key* is not in *d*.

`len(d)`
Number of key-value pairs in *d*.

`key in d`
`True` if *key* has an entry in *d*; otherwise, `False`.

`key not in d`
`True` if *key* does not have an entry in *d*; otherwise, `False`.

`del d[key]`
Delete entry for *key* in *d*. Raises `KeyError` if *key* is not in *d*.

`sorted(d)`
Return sorted list of keys in *d*. Use `sorted(d, key=d.get)` to sort the keys by value.

Table 3.22 lists some of the basic operations for a dictionary *d*. The first two are the most important, to set and retrieve `d[key]`. Dictionaries are designed so these two operations are very fast.

Example: Using the `status` dictionary above, `status[200]` is `"ok"`.

Example: `301 in status` is false.

Example: Using the `population` dictionary above,

    population["Rome"] += 0.11

increases the entry for Rome to 2.98.

⟹ Caution: Attempting to access `d[key]` before a value has been set for *key* is an error.

Example: `population["Madrid"] += 0.08` raises a `KeyError`

The reason is that the right-hand side of the full assignment is executed first:

    population["Madrid"] = population["Madrid"] + 0.08

and there is no entry for Madrid.

Dictionary Methods

Dictionaries are objects and therefore also have their own methods, some of which are listed in Table 3.23.

TABLE 3.23 Dictionary methods

`d.get(key, default)`
Return value for *key* if *key* in *d*; otherwise return optional *default* (or `None` if no *default* given).

`d.pop(key, default)`
Delete and return value for *key* if *key* in *d*; otherwise return optional *default*. Raises `KeyError` if *key* not in *d* and no *default*.

`d.keys()`
View of keys in *d*.

`d.values()`
View of values in *d*.

`d.items()`
View of (`key, value`) pairs in *d*.

⟶ Note: The `get()` method is similar to using `d[key]`, except that it returns a default value instead of raising an error when the *key* is not present.

The `keys()`, `values()`, and `items()` methods return special dictionary *views* that update as the contents of the dictionary changes. The `list()` type converter can convert a view to a list if you need it.

Dictionary Loops

It is easy to loop over the keys in a dictionary *d* with a **for** loop:

```
for key in d:
    body
```

Remember that in a dictionary, there is no particular order to the keys. Use the **sorted()** function if you need sorted keys.

Type Considerations

Because they are used for lookup, dictionary keys must be immutable. Thus, for example, lists may not be used as dictionary keys. Values, on the other hand, may be of any type. There is also no requirement that all keys have the same type, so it is possible to mix, for example, integer and string keys. Dictionaries themselves are mutable objects and have type **dict**.

EXERCISES

1. Which of the following types may be used as keys in a Python dictionary: integer, float, string, list, boolean, dictionary? Explain your answers.

2. Which of the following types may be used as values in a Python dictionary: integer, float, string, list, boolean, dictionary? Explain your answers.

3. Write a Python statement to create a dictionary named `rating`, where each key is a movie title and the value is your rating of that film from 1 (low) to 5 (high). Include at least four entries.

4. Write a Python statement to create a dictionary named `friend`, where each key is a person's name and the value is a list of their friends' names. Include at least four entries.

5. Using the `population` dictionary on page 136, do the following:

   (a) Write statements to add three new entries (without re-creating the dictionary).

   (b) Write a statement to delete the entry for New York.

   (c) Write an expression to test to see if there is an entry for Tokyo.

   (d) Give the value of: `list(population.keys())`.

6. Using the `status` dictionary on page 136, do the following:

   (a) Write statements to add three new entries (without re-creating the dictionary).

   (b) Write a statement to change the entry for status 200 to "OK."

   (c) Write an expression that gives the number of entries in `status`.

   (d) Give the value and explain the effect of: `status.pop(404)`.

7. Explain why this simpler version of the loop in Example 3.8 without an `if` statement does not work:

```
for item in items:
    count[item] += 1        # Error!
```

8. Rewrite the `frequency()` function in Example 3.8 to remove the `if` statement by using `get()` with a default value. Include a program to test your function.

9. Incorporate the `frequency()` function of Example 3.8 into a complete program to determine the word frequencies in a text file. Caution: large amounts of output can be very slow in IDLE.

10. Write a function `letter_frequency(text)` to return the frequencies of the letters (only) in the string *text*, ignoring upper- and lowercase. Include a program to test your function. Letter frequencies are important for encryption, compression, modern keyboard designs, and some games.

11. Write a function `char_frequency(text)` to return the frequencies of every character in the string *text*. Include a program to test your function.

12. Write a function `forms(verb)` to return the present indicative conjugated forms of *verb* for any regular -ar, -er, or -ir Spanish verb. Use a Python dictionary to store the list of endings for each family of verbs, and include a program to test your function.

13. Rewrite the `complementary_base(base)` function of Example 3.2 to look up the complement of the given *base* in a Python dictionary.

14. Write a function `top(n, items)` to return a list of the top *n* elements in the list *items*, based on frequency. Include a program to test your function.

15. Write a `mode(items)` function to return the mode of the values in the list *items*. The **mode** of a data set is the value that occurs most often; if there is a tie, return any one of the items with the highest count. Include a program to test your function.

16. Write a program to find the word(s) with the most anagrams from a text dictionary file. Hint: use a Python dictionary to count how many words have the same string of sorted characters.

17. Write a program to provide an interactive dictionary, where a user can add, change, and look up definitions. The program starts with an empty dictionary, and then entries are gradually created by the user. The start of an interactive session might look like this, with the user's input underlined:

```
Welcome to PyDict
        [a]dd or change a definition
        [l]ookup a word
        [q]uit
Your choice: l
Word to look up: python
Not in dictionary
        [a]dd or change a definition
        [l]ookup a word
        [q]uit
Your choice: a
Word: python
Definition: A long, large snake.
        [a]dd or change a definition
        [l]ookup a word
        [q]uit
Your choice: l
Word to look up: python
Definition: A long, large snake.
        [a]dd or change a definition
        [l]ookup a word
        [q]uit
```

PROJECT: ELIZA

In the mid-1960s, Joseph Weizenbaum at MIT wrote ELIZA, a ground-breaking program in **artificial intelligence** that allowed users to converse with it using plain English. ELIZA was meant to sound like a Rogerian therapist: friendly, non-judgmental, and reflective, using questions to get the user to talk further about him or herself. By searching for certain keywords, ELIZA could provide very focused responses, in addition to using a variety of generic statements and questions. These ideas are now commonplace, thanks to the widespread use of virtual assistants.

The following exercises outline one way to design a program like ELIZA.

EXERCISES

1. Begin your own version of ELIZA by writing a program to have a conversation with the user in which your program always says the same thing. Continue the conversation until the user says, "Bye."

2. Write a function `random_response()` to return a random response that fits the personality of your ELIZA. Modify your program to use it instead of always saying the same thing.

3. Write a function `respond(text)` to return a more appropriate response based on what the user has said in *text*. Set up a keyword-response dictionary so that if the user types one of the keywords within *text*, a more targeted response can be given. Return a random response if none of the keywords match.

4. Ask the user for his or her name, and incorporate their name into some of the program's responses, both random and keyword based.

5. Make adjustments so that important terms such as keywords or "Bye" are found regardless of punctuation or upper-/lowercase.

6. Research the Turing Test. Report your findings, including its connection with ELIZA.

PROJECT: READING DNA FRAMES

Exercise 12 in Section 3.2 describes (without the chemistry) how sense DNA is converted to mRNA through transcription. After transcription, mRNA is converted to a sequence of amino acids through a process called **translation**. RNA translation takes groups of three nucleotide bases called **codons**, and converts each codon to an amino acid according to Table 3.24.

Table 3.24 uses standard one-letter abbreviations for the 20 amino acids. A few codons signal a STOP in the sequence; these are indicated by _'s in the table.

TABLE 3.24 Amino acid translation

		Base 2			
Base 1	U	C	A	G	Base 3
U	F	S	Y	C	U
	F	S	Y	C	C
	L	S	_	_	A
	L	S	_	W	G
C	L	P	H	R	U
	L	P	H	R	C
	L	P	Q	R	A
	L	P	Q	R	G
A	I	T	N	S	U
	I	T	N	S	C
	I	T	K	R	A
	M	T	K	R	G
G	V	A	D	G	U
	V	A	D	G	C
	V	A	E	G	A
	V	A	E	G	G

Example: Look up AGC in Table 3.24.

Find A in the first column, then look down the G column (as the second base), and find the C row using the last column. This gives the amino acid S, which is serine.

Reading Frames

Proteins are encoded in this way as strings of amino acids. Given a strand of RNA, there is no way to know ahead of time where to start reading the 3-base codons. Thus, we need to try all three possibilities to find possible proteins: starting at the first base (index 0), second base (index 1), or third (index 2). Each of these is called a **reading frame. Open reading frames** are reading frames that have long stretches without STOP codons, and these are good candidates for protein-coding sequences.

Given a strand of DNA, as opposed to RNA, remember that there are two strands bound together in a double-helix: the given DNA strand and its reverse complement. Thus, to search a segment of DNA for protein sequences requires checking *six* reading frames: three based on the given strand, and three from its reverse complement.

FASTA Data Files

FASTA is an text-based data file format for DNA sequences. Each sequence in a file starts with a header line that begins with a >. The lines that follow, up to the next header line, contain the DNA sequence.

EXERCISES

1. Write a function `translate(rna, frame)` to translate one frame of *rna* to the corresponding sequence of amino acids. The parameter *frame* = 0, 1, or 2 specifies which frame to translate. Use a Python dictionary to represent Table 3.24.

2. Write a program to process a FASTA file containing information for only one DNA sequence. Include these functions:

   (a) `first_line(text)` to return the first line of text from the FASTA file, which is the header.

   (b) `other_lines(text)` to return a single uppercase string containing the DNA sequence that follows the first-line header.

   (c) `print_frames(rna)` to print translations of the three reading frames of *rna*. Convert `STOP` codons to newlines (\n) so that proteins are easier to find.

   Print the header followed by translations of all six reading frames.

3. Write a program to process FASTA files containing more than one DNA sequence.

CHAPTER 4

Images

4.1   CREATING IMAGES

Digital images are everywhere, and while many software packages allow you to create and modify images, learning to manipulate images in Python provides insight into how many standard effects work and allows you to implement your own creative ideas.

Pillow: The Python Imaging Library (PIL)

The programs in this chapter use the Pillow implementation of **PIL**, the Python Imaging Library (`https://python-pillow.org/`). Follow the online instructions to install the latest version of Pillow on your system. The examples here were written with Pillow version 4.2.1.

Example 4.1 shows how to create an image of the national flag of the Netherlands with PIL. The Dutch flag consists of three equal horizontal stripes of bright vermillion, white, and cobalt blue.

Pixel Grids

Images are usually represented in software as two-dimensional grids of **pixels**, where each pixel has a specified color. Pixel locations are given by (i, j) coordinates with $(0, 0)$ at the upper left corner, i coordinates increasing to the right, and j coordinates increasing downward. For example, the pixel coordinates for a tiny 5×5 image are shown in Fig. 4.1. Notice the differences between these coordinates and mathematical graphs in the xy-plane: the origin is in the upper left corner instead of lower left, and the j coordinate grows down instead of up.

RGB Color

There are many ways to represent color in software; for now, the most common is RGB. Each pixel in a 24-bit **RGB image** has one byte of storage for each of

```
1   # dutchflag.py
2
3   from PIL import Image
4
5   def dutchflag(width, height):
6       """Return new image of Dutch flag."""
7       img = Image.new("RGB", (width, height))
8       for j in range(height):
9           for i in range(width):
10              if j < height/3:
11                  img.putpixel((i, j), (174, 28, 40))
12              elif j < 2*height/3:
13                  img.putpixel((i, j), (255, 255, 255))
14              else:
15                  img.putpixel((i, j), (33, 70, 139))
16      return img
17
18  def main():
19      img = dutchflag(600, 400)
20      img.save("dutchflag.jpg")
21
22  main()
```

Example 4.1 Dutch flag.

three components: red (R), green (G), and blue (B). Each byte is interpreted as an unsigned integer, so its value is between 0 and 255. At 3 bytes per pixel, image files become large very quickly, so they are usually compressed.

RGB values represent the intensities of red, green, and blue wavelengths of light, and so $(0, 0, 0)$ is the darkest color (black) because each wavelength has no intensity, while $(255, 255, 255)$ is the brightest (white) because each wavelength is at full intensity. Other common RGB colors are listed in Table 4.1.

Example: RGB $(33, 70, 139)$ is the cobalt blue used in the Dutch flag.

$\longrightarrow$ Note: Mixing colors of light is different than mixing paint. Light colors are additive, whereas paint colors are subtractive.

Python Tuples

Python programs represent both pixel coordinates and RGB values as integer tuples. A **tuple** is a sequence like a list, except tuple objects are immutable. Tuples are written with parentheses instead of square brackets: for example,

$$i \rightarrow$$

$(0,0)$	$(1,0)$	$(2,0)$	$(3,0)$	$(4,0)$
$(0,1)$	$(1,1)$	$(2,1)$	$(3,1)$	$(4,1)$
$(0,2)$	$(1,2)$	$(2,2)$	$(3,2)$	$(4,2)$
$(0,3)$	$(1,3)$	$(2,3)$	$(3,3)$	$(4,3)$
$(0,4)$	$(1,4)$	$(2,4)$	$(3,4)$	$(4,4)$

$j \downarrow$

Figure 4.1   5x5 pixel grid.

TABLE 4.1   Basic RGB colors

Color	RGB Value
black	(0, 0, 0)
red	(255, 0, 0)
green	(0, 255, 0)
blue	(0, 0, 255)
yellow	(255, 255, 0)
magenta	(255, 0, 255)
cyan	(0, 255, 255)
white	(255, 255, 255)

t = (1, 2, 3) is a tuple, whereas u = [1, 2, 3] is a list. Tuples support all sequence operations, such as indexing and slicing, but because they are immutable, they do not support list assignment or mutator methods. The tuple type is **tuple**, and its type converter function is **tuple()**.

Tuples are sometimes referred to by their size; for example, (1, 7) is a **2-tuple** and (0, 100, 0) is a **3-tuple**.

Tuples may be **unpacked** into their separate components with an assignment statement:

```
var1, var2, ..., varN = tuple
```

Each variable on the left is assigned to the corresponding value from the *tuple* on the right, as long as the number of items match.

Example: r, g, b = (174, 28, 40)

This unpacking statement assigns r = 174, g = 28, and b = 40.

PIL Image Objects

Example 4.1 shows how to create an image with PIL. The `dutchflag()` function creates a PIL **image object** named `img` and then sets each pixel in `img` to the appropriate color by calling `img.putpixel()`. Then, the image object returned by `dutchflag()` is saved to a JPEG file in `main()`.

Table 4.2 lists the two main ways of obtaining an image object, either for a new image or from an existing image file.

TABLE 4.2 **PIL** module: get image objects

`Image.new(mode, size, color)`
Return new image object. The *mode* is a string (use `"RGB"`), *size* is an integer tuple (`width, height`), and *color* is an integer tuple (`r, g, b`). The default color is black.
`Image.open(filename)`
Return image object for existing file *filename*. Raises `OSError` on errors opening or reading file; closes file automatically.

Given an image object named *img* returned from either `Image.new()` or `Image.open()`, Table 4.3 lists the most important methods that can be called on it.

TABLE 4.3 **PIL** module: basic image methods

`img.putpixel(location, color)`
Set pixel at *location* to *color*. The *location* is an integer tuple (`i, j`), and *color* is an integer tuple (`r, g, b`).
`img.getpixel(location)`
Get RGB tuple at pixel *location*. The *location* is an integer tuple (`i, j`).
`img.save(filename)`
Save image to *filename*. Raises `OSError` on errors while saving. Raises `KeyError` if filetype cannot be inferred from *filename* extension.

$\longrightarrow$ Note: Pay attention to where tuples are expected in Tables 4.2 and 4.3, because it is easy to forget the parentheses that must be around a tuple.

Image Loops

Example 4.1 contains a set of nested **for** loops in lines 8–9:

```
for j in range(height):
    for i in range(width):
```

You may wonder about the order of these loops—why is *j* in the outer loop instead of *i*? Looping over *j* first isn't necessary, but it is common because image data is stored by rows, starting at the top. All of the $j = 0$ pixels occur first, then all of the $j = 1$ pixels, and so on:

$(0,0)$	$(1,0)$	$(2,0)$	$(3,0)$	$(4,0)$
$(0,1)$	$(1,1)$	$(2,1)$	$(3,1)$	$(4,1)$

Putting j in the outer loop visits each of the pixels in the same order they are stored.

EXERCISES

1. Sketch a 6x6 pixel grid and show where the pixels $(0,4)$, $(2,2)$, and $(5,1)$ are in the grid.

2. Sketch a 6x6 pixel grid and show where the pixels $(5,3)$, $(2,0)$, and $(1,4)$ are in the grid.

3. Describe the difference between RGB $(100,0,0)$ and RGB $(0,0,100)$.

4. Describe the difference between RGB $(20,0,0)$ and RGB $(200,0,0)$.

5. Describe the color RGB $(30,30,30)$.

6. Describe the color RGB $(220,220,220)$.

7. Suppose `color` is an RGB tuple. Describe what `color[0]` is.

8. Suppose `color` is an RGB tuple. Describe what `color[2]` is.

9. Can tuples be used as dictionary keys? Explain why or why not.

10. Write a Python statement to unpack the tuple named `pixel` into separate i and j coordinates.

11. Sketch the image produced by this code:

```
for j in range(200):
    for i in range(100):
        if j < 50:
            img.putpixel((i, j), (150, 0, 0))
        elif i > 50:
            img.putpixel((i, j), (0, 0, 150))
        else:
            img.putpixel((i, j), (255, 255, 255))
```

12. Sketch the image produced by this code:

```
for j in range(100):
    for i in range(200):
        if i < 50:
            img.putpixel((i, j), (150, 0, 0))
        elif j > 50:
            img.putpixel((i, j), (0, 0, 150))
        else:
            img.putpixel((i, j), (255, 255, 255))
```

13. Modify Example 4.1 to save the flag image as a PNG (portable network graphics) file.

14. Modify Example 4.1 to handle an OSError when saving.

15. Write a randomcolor() function to return a random RGB color tuple. Test your function in an image.

16. Write a randomgray() function to return a random shade of gray. Test your function in an image.

17. Write a randomflag(width, height) function to return the image of a flag like the Dutch flag except that the top and bottom stripes are solid colors chosen at random.

18. Write a randomstripe(width, height) function to return the image of a white flag with a horizontal stripe made up of random pixel colors in the middle third of the flag.

19. Write an italianflag(width, height) function to return the image of the flag of Italy with correct proportions and colors.

20. Write an frenchflag(width, height) function to return the image of the flag of France with correct proportions and colors.

21. Write a finnishflag(width, height) function to return the image of the flag of Finland with correct proportions and colors. (Approximate the shade of blue, since it is not possible to represent as RGB.)

22. Write a swedishflag(width, height) function to return the image of the flag of Sweden with correct proportions and colors.

23. Write an rgbflag(width, height) function to return an image with equal horizontal stripes of bright red $(255, 0, 0)$, green $(0, 255, 0)$, and blue $(0, 0, 255)$.

4.2 COLOR TRANSFORMATIONS

Many common image effects involve changing pixels based only on the color of the original pixel. Example 4.2 converts any existing image to **grayscale**, or what we usually refer to as "black-and-white" as opposed to color. It uses the `Image.open()` function from Table 4.2 and the `getpixel()` method from Table 4.3, both described in the previous section.

$\longrightarrow$ Note: Use any of your own images for programs like this. Adjust the size of the image if programs take too long to run.

```python
1  # grayscale.py
2
3  from PIL import Image
4
5  def grayscale(img):
6      """Return copy of img in grayscale."""
7      width, height = img.size
8      newimg = Image.new("RGB", (width, height))
9      for j in range(height):
10         for i in range(width):
11             r, g, b = img.getpixel((i, j))
12             avg = (r + g + b) // 3
13             newimg.putpixel((i, j), (avg, avg, avg))
14     return newimg
15
16 def main():
17     img = Image.open("lake.jpg")
18     newimg = grayscale(img)
19     newimg.save("lake_gray.jpg")
20
21 main()
```

Example 4.2 Grayscale.

In order to loop over the pixels in existing images, we need to know the image's size. PIL provides the size as an attribute of the image object.

Syntax: Data Attributes

Every object in Python stores its data in variables called **attributes** of the object. Because objects are referred to as instances, data attributes are also known as **instance variables**, and in some cases they are called **fields**. Most of the time, we access and manipulate data attributes through an object's methods, but some objects, especially in Python, are designed to have data directly available in instance variables.

Data attributes are accessed with the same dot notation used for method calls:

```
object.attribute
```

This is exactly the same as a method call (see page 111), except there are *no parentheses* after the name of a data attribute. Similar syntax is used because methods are also considered to be attributes of the object.

PIL image objects have a `size` attribute, described in Table 4.4. Line 7 of Example 4.2 unpacks the `img.size` tuple into separate `width` and `height` variables.

TABLE 4.4 **PIL** module: image attributes

`img.size` Size of *img* in pixels, as (`width,height`) tuple.
`img.width` Width of *img* in pixels.
`img.height` Height of *img* in pixels.

Images Are Mutable

The `putpixel()` method certainly qualifies as a mutator, implying that images are a mutable type. Notice, though, that the `grayscale()` function in Example 4.2 does not modify its parameter *img*: instead, it returns a copy of the image in grayscale.

An alternative design would be to write a `to_grayscale(img)` function that modifies *img* rather than returning a new image. That is an equally reasonable choice for this task, but in the long run, returning a modified copy is more flexible. It is also necessary for many of the other types of transformations we will consider.

Relative Luminance

You may notice that grayscale images produced by Example 4.2 do not always look quite right. (If you have not noticed any problem, see Exercise 9.) One reason is that the average of the RGB values at a pixel does not give an accurate estimate of its relative brightness.

The physical basis of RGB color is that human eyes have three types of cones that respond to different wavelengths of color: one centered around red, one green, and one blue. These cones have noticeably different sensitivities, meaning that our eyes are much more sensitive to green light than red, and more sensitive to red light than blue.

Relative **luminance** is an attempt to estimate the brightness of a color as perceived by our eyes. One rough approximation for the luminance of a color (r, g, b) is:

$$0.2 * r + 0.7 * g + 0.1 * b$$

The coefficients 0.2, 0.7, and 0.1 add up to 1.0, so that the result of this calculation stays between 0 and 255.

⟹ Caution: As in Example 4.2, be careful to always provide integer RGB tuples to `putpixel()`; otherwise, you will get a `TypeError` exception.

EXERCISES

1. Explain all of the parentheses in the calls to `getpixel()` and `putpixel()` in Example 4.2.

2. Explain why integer division `//` is used to calculate the average in Example 4.2. Try changing it to division `/`, and explain the outcome.

3. Describe the effect of this code on an image:

```
for j in range(height):
    for i in range(width):
        r, g, b = img.getpixel((i, j))
        img.putpixel((i, j), (r, 0, b))
```

4. Describe the effect of this code on an image:

```
for j in range(height):
    for i in range(width):
        r, g, b = img.getpixel((i, j))
        img.putpixel((i, j), (g, b, r))
```

5. Rewrite Example 4.2 without unpacking `img.size`.

6. Modify Example 4.2 to compute the average without unpacking the `(r, g, b)` color tuple from `getpixel()`.

7. Modify Example 4.2 to handle any `OSErrors`.

8. Write a `to_grayscale(img)` function to modify *img* to be grayscale, instead of returning a modified copy. Include a program to test your function.

9. Try Example 4.2 on the RGB flag of Exercise 23 in the previous section. Describe and explain the result. Does it look right?

10. Write a `grayscale_lum(img)` function to return a copy of *img* in grayscale based on relative luminance instead of the average RGB. Include a separate function to return the luminance of an RGB color and a program to test your function.

11. Write a `negative(img)` function to return a copy of *img* that is its negative. Include a separate function to return the negative of a color (r, g, b), which is $(255 - r, 255 - g, 255 - b)$, as well as a program to test your function.

12. Write a `twotone(img, dark, light)` function to return a copy of *img* in which all dark pixels are set to the color *dark* and all light pixels are set to *light*. Use a luminance function to decide between dark and light, and include a program to test your function.

13. Write a `darken(img)` function to return a darkened copy of *img* by decreasing each RGB component by ten percent. Include a separate `decrease(component)` function to do the calculation on one component, being careful to return a valid integer. Test your function.

14. Write a `lighten(img)` function to return a lightened copy of *img* by increasing each RGB component by ten percent. Include a separate `increase(component)` function to do the calculation on one component, being careful to return a valid integer between 0 and 255. Test your function.

PROJECT: GAMMA CORRECTION

Calculating the luminance of pixels in JPEG files is actually more complicated than implied in the previous section. The reason is that JPEG pixels have a gamma correction built into them that affects computations based on those pixel values. **Gamma correction** takes into account how display technologies work so that color values between 0 and 255 appear to increase linearly. The issue is that display brightness does not increase linearly, so gamma correction compensates for that.

Because JPEG pixels are already corrected this way, that correction needs to be removed before computing luminance, and then reapplied before storing the new pixel color. A rough approximation to remove the correction is to square the value. Then reapplying the correction can be done with a square root. This gives a new calculation for luminance of

$$\sqrt{0.2 * r^2 + 0.7 * g^2 + 0.1 * b^2}.$$

Float RGB

However, there is one more complication: the use of squaring and square roots assumes that the RGB values are floats in the interval $[0, 1]$ instead of integers between 0 and 255. That means we need to convert the RGB components to floats before computing luminance, and then convert the result back to an integer to create the gray RGB color.

The following exercises suggest one way to organize these computations.

EXERCISES

1. Write a function rgb_to_floats(*color*) to return an RGB tuple of floats in $[0, 1]$ that corresponds to the integer RGB tuple named *color*.

2. Write an adjusted_brightness(*color*) to compute the luminance of a *color* of $[0, 1]$ floats using the formula above that is adjusted to take into account gamma correction.

3. Write a float_to_int(*value*) function to return an integer between 0 and 255 corresponding to the float *value* in $[0, 1]$.

4. Write a luminance_gray(*color*) to return a gray integer RGB tuple corresponding to the relative luminance of the given integer RGB *color*.

5. Write a grayscale_lum(*img*) function to return a copy of *img* in grayscale based on relative luminance, adjusted to take into account gamma correction. Compare the results of this function to using luminance without gamma correction on the RGB flag (see Exercise 9 in Section 4.2).

PROJECT: COLOR QUANTIZATION

Another interesting effect that uses float RGB values in $[0, 1]$ (see the previous project) is color quantization. **Quantization** reduces the number of different colors in the image—which does not sound good—but it can be used to create quite interesting and dramatic images.

One way to reduce the number of colors is to reduce each R, G, B component to one of a small number of possible values. Given a float in $[0, 1]$ and an integer n, the following steps will produce one of $n + 1$ different values. An example with $n = 5$ is shown to illustrate what happens:

1. Multiply by n. The result is a float in $[0, 5]$.

2. Add 0.5. The result is a float in $[0.5, 5.5]$.

3. Take the integer part (floor). This gives one of $0, 1, 2, 3, 4, 5$.

4. Divide by n. The result is one of $0, \frac{1}{5}, \frac{2}{5}, \frac{3}{5}, \frac{4}{5}, 1$.

Since we need an integer RGB at the end, the last step can be replaced with

4′. Multiply by 255 and integer-divide by n. The result is one of 0, 51, 102, 153, 204, 255.

EXERCISES

1. Determine the number of different RGB colors that may result from using steps 1–4′ above with $n = 3$ in each component.

2. Write a function rgb_to_floats(*color*) to return an RGB tuple of floats in $[0, 1]$ that corresponds to the integer RGB tuple named *color*.

3. Write a function quantize_component(*value*, n) to return a quantized color component by performing the computation outlined in steps 1–4′ above. Given a float *value* in $[0, 1]$ and integer n, return one of $n + 1$ integers between 0 and 255.

4. Write a function quantize_rgb(*color*, n) to return a quantized RGB tuple given an integer RGB *color* and integer n.

5. Write a function quantize(*img*, n) to return a quantized copy of *img* using n color levels.

4.3   SIZE TRANSFORMATIONS

Changing an image's size always requires creating a new image object. Although we have access to an image's size attribute, that value should only be viewed, not changed, because it has to match the rest of the underlying pixel data.

Example 4.3 shows one way to create a copy of an image that doubles the width and height. It does not require any new library functions.

```
 1  def double(img):
 2      """Return copy of img with doubled width, height."""
 3      width, height = img.size
 4      newimg = Image.new("RGB", (2*width, 2*height))
 5      for j in range(height):
 6          for i in range(width):
 7              color = img.getpixel((i, j))
 8              newimg.putpixel((2*i, 2*j), color)
 9              newimg.putpixel((2*i, 2*j + 1), color)
10              newimg.putpixel((2*i + 1, 2*j), color)
11              newimg.putpixel((2*i + 1, 2*j + 1), color)
12      return newimg
```

Example 4.3   Double image width and height.

Choosing Loops

The first choice to make when working with images of different sizes is which image to loop over, the original or the new image. Usually, both ways work, and the choice is a matter of preference. The `double()` function of Example 4.3 loops over the width and height of the original *img*.

That choice determines the next question to ask: starting with a pixel in the original image (the one we loop over), what has to happen in the new image? In this function, the new image will need four pixels for each pixel in the original:

> *for each original pixel:*
> *get its color*
> *put that color in four new pixels*

The next step then is to determine the coordinates of the four new pixels based on the coordinates of the original pixel.

Calculating Coordinates

Because we loop over the original image with variables i and j in lines 5 and 6, we need to get from having original-image coordinates (i, j) to corresponding pixel coordinates in the new image:

$$\text{Original } (i, j) \longrightarrow \text{New } (?, ?)$$

To do this, draw pictures, write out specific examples, and look for patterns. For example,

$$(0, 0) \longrightarrow (0, 0), (1, 0), (0, 1), (1, 1)$$
$$(2, 0) \longrightarrow (4, 0), (5, 0), (4, 1), (5, 1)$$
$$(1, 3) \longrightarrow (2, 6), (3, 6), (2, 7), (3, 7)$$
$$\cdots$$

Then, in this case, focus on one family of new pixels at a time, and try to determine a formula for those coordinates. For example, the upper left corners, listed first above, all match the pattern $(2i, 2j)$.

Looping over the New Image

The same process works if you choose instead to loop over the new image, but the individual steps may look very different because they will be New $\longrightarrow$ Old. Start by deciding what to do with each new pixel:

> *for each new pixel:*
> *....*

Exercise 1 asks you to finish this approach.

EXERCISES

1. Rewrite the double(*img*) function to use a loop over the new image instead of the original *img*. Include a program to test your function.

2. Write a function half(*img*) to return a copy of *img* half as wide and half as tall as the original. Loop over the new image, and scale down by choosing one pixel from each 2 × 2 block. Include a program to test your function.

3. Write a function half(*img*) to return a copy of *img* half as wide and half as tall as the original. Loop over the old image, and scale down by choosing one pixel from each 2 × 2 block. Test your function with images that have at least one odd side length.

4. Write a blur(*img*) function that works by first scaling down by half, and then doubling the result. Test your function.

5. Describe the effect of half(double(*img*)).

6. Write a function scale_up(*img*, *n*) to return a copy of *img* that is *n* times as wide and *n* times as tall as the original. Assume *n* is an integer, and test your function.

7. Write a function scale_down(*img*, *n*) to return a copy of *img* that is $1/n$ times as wide and $1/n$ times as tall as the original. Assume *n* is an integer. Scale down by choosing one pixel from each *n* × *n* block. Include a program to test your function.

8. Write a function scale_down_avg(*img*, *n*) to return a copy of *img* that is $1/n$ times as wide and $1/n$ times as tall as the original. Assume *n* is an integer. Scale down by averaging the colors of all pixels in each *n* × *n* block. Compare the results with choosing one pixel from each block.

9. Write a function pixelate(*img*, *n*) to return a copy of *img* that is first scaled down and then scaled back up to the original size. Include a program to test your function.

10. Write a function tile(*img*, *n*) to return a new image containing n^2 full-size copies of *img* tiled to make an *n* × *n* square. Include a program to test your function.

11. Write a function albumcover(*img*, *n*) that combines tiling with scaling down to return a tiled copy of *img* that is the same size as the original. Include a program to test your function.

12. Write a function crop(*img*, *upper_left*, *lower_right*) to return a new image consisting of the cropped rectangle in *img* defined by the two points given as parameters. Include a program to test your function.

4.4 GEOMETRIC TRANSFORMATIONS

Geometric transformations include rotations, reflections, and mirror images. They are similar to size transformations, although there is usually no advantage to looping over the new image, so you can always plan to use the original. Example 4.4 demonstrates how to rotate an image 90° to the right.

```
1  def rotate_right(img):
2      """Return copy of img rotated right 90 degrees."""
3      width, height = img.size
4      newimg = Image.new("RGB", (height, width))
5      for j in range(height):
6          for i in range(width):
7              color = img.getpixel((i, j))
8              newimg.putpixel((height - j - 1, i), color)
9      return newimg
```

Example 4.4 Rotate image.

Calculating Coordinates

The main challenge with these transformations is in determining where to put the new pixel, assuming that loops are over the original image. Drawing a picture such as Fig. 4.2 can help. The dashed arrows show the new coordinates within the rotated image that need to be calculated:

$$\text{new } i = height - j - 1$$
$$\text{new } j = i$$

These values are used in line 8 of Example 4.4.

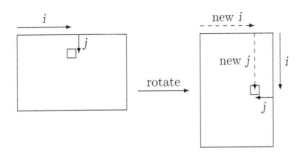

Figure 4.2 Coordinates of rotated pixel.

EXERCISES

1. What is important to notice in line 4 of Example 4.4?

2. Draw a diagram similar to Fig. 4.2 for a 90-degree rotation to the left. Use it to determine the new coordinates of the pixel in the rotated image.

3. Write a function rotate_left(img) to return a copy of img rotated 90 degrees to the left. Include a program to test your function.

4. Write a function rotate_180(img) to return a copy of img rotated 180 degrees. Include a program to test your function.

5. Write a function flip_top_bottom(img) to return a copy of img reflected across a horizontal line through the center. Include a program to test your function.

6. Write a function flip_left_right(img) to return a copy of img reflected across a vertical line through the center. Include a program to test your function.

7. Write a function mirror_top_to_bottom(img) to return a copy of img with the top half mirrored onto the bottom half. Include a program to test your function.

8. Write a function mirror_bottom_to_top(img) to return a copy of img with the bottom half mirrored onto the top half. Include a program to test your function.

9. Write a function mirror_left_to_right(img) to return a copy of img with the left half mirrored onto the right half. Include a program to test your function.

10. Write a function mirror_right_to_left(img) to return a copy of img with the right half mirrored onto the left half. Include a program to test your function.

11. Write a function transpose(img) to return a copy of img with its coordinates reversed, (i, j) to (j, i). This has the effect of reflecting across the diagonal through $(0, 0)$. Include a program to test your function.

4.5   COMBINING IMAGES

So far, all of our image programs have created or manipulated one image at a time. In this section, we look at different ways of combining images, starting with blending two images in Example 4.5. No new library functions are required, but this program illustrates a couple of new techniques for handling errors.

```
1   # blend.py
2
3   from PIL import Image
4
5   def avg(x, y):
6       """Return average of x and y."""
7       return (x + y)/2
8
9   def blend(img1, img2):
10      """Return blend of two images of the same size."""
11      if img1.size != img2.size:
12          raise ValueError("Images are not the same size.")
13      width, height = img1.size
14      newimg = Image.new("RGB", (width, height))
15      for j in range(height):
16          for i in range(width):
17              r1, g1, b1 = img1.getpixel((i, j))
18              r2, g2, b2 = img2.getpixel((i, j))
19              newimg.putpixel((i, j),
20                              (int(avg(r1, r2)),
21                               int(avg(g1, g2)),
22                               int(avg(b1, b2))))
23      return newimg
24
25  def main():
26      try:
27          img1 = Image.open("lake.jpg")
28          img2 = Image.open("lake2.jpg")
29          newimg = blend(img1, img2)
30          newimg.save("lake_blend.jpg")
31      except (OSError, ValueError) as err:
32          print("Error:", err)
33          print("Unable to process blending.")
34
35  main()
```

Example 4.5 Blend images.

Raising Exceptions

In order to blend two images, they need to be the same size. If the code for `blend()` makes that assumption but images of different sizes are passed in, then an exception will be thrown when an attempt is made to access a pixel that doesn't exist. Instead of waiting for that predictable and bad situation to happen, lines 11–12 of Example 4.5 check the sizes and raise an exception to explain the error. This will more clearly describe the problem to the caller than a pixel access error.

Whenever possible, use standard Python exception types instead of creating your own. Refer back to Table 3.19 on page 132 for a list of common errors.

Syntax: Handling Different Exceptions

Given that `blend()` may raise a `ValueError` exception because of mismatched sizes, several different things could go wrong running Example 4.5. Line 31 shows how to catch more than one exception type with the same **except** clause:

```
except (ExceptionType1, ExceptionType2, ...) as errorname:
    errorbody
```

In the case of Example 4.5, it is appropriate to treat these errors together because any of them result in the program not being able to run.

Errors that need to be handled differently can be put into separate **except** clauses.

EXERCISES

1. Write a `mix(x, y, p)` function to return $px + (1 - p)y$, a more general mixing of x and y than average, with an adjustable proportion p. For example, setting $p = 0.3$ mixes 30% x and 70% y. Use `mix()` to write a `blend(img1, img2, p)` function to return a blending of *img1* with *img2* using fraction p. Set a default value of $p = 0.5$. Include a program to test your functions.

2. Write a `paste(source, destination, upper_left)` function to return a copy of the *destination* image with the *source* image pasted into it, starting at the *upper_left* corner in *destination*. Paste only as much of *source* as will fit; there should not be an error if it does not fit entirely inside *destination*. Include a program to test your function.

3. Write a `side_by_side(left, right)` function to return a new image consisting of the *left* and *right* images next to each other, side-by-side, without any scaling. The two images must have the same height. Check for errors in `main()`.

4. Write an `above_below(top, bottom)` function to return a new image consisting of the *top* image above the *bottom*, without any scaling. The two images must have the same width. Check for errors in `main()`.

PROJECT: IMAGE FILTERS

Many common image manipulations involve the application of a filter. An **image filter** is a square array of coefficients that describes how old pixel values will be combined to find a new pixel value. Each entry in the array is a coefficient to be multiplied by the color in that location. The resulting RGB values are then added together to produce the final result. The center of the filter corresponds to the position of the pixel whose color is currently being computed.

For example, this is a 3×3 filter that takes 0 times the current color in the center plus $1/8$ times each of the colors in the surrounding pixels:

1/8	1/8	1/8
1/8	0	1/8
1/8	1/8	1/8

The fraction $1/8$ is used so that the resulting image has approximately the same brightness as the original, because the sum of the pixel values from the filter equals one "pixel's worth" of color. The effect of this filter is to blur the original image since the (possibly) distinctive color that used to be in the center location has been replaced by a combination of all the colors around it.

The following filter performs a type of edge detection by highlighting whatever is different in the center pixel while removing the colors that surround it:

0	−1	0
−1	4	−1
0	−1	0

It gives a combined zero pixel's worth of color, which means that the resulting image will be quite dark. Either integers or fractions may be used in filters, and coefficients may be positive, negative, or zero.

Applying Filters

In order to apply a filter to an image, the **for** loops that iterate over all pixels in the original image must be adjusted to avoid pixels on the edges. This is because we need to be able to access all of the pixels in the 3×3 grid surrounding any pixel to which we apply the filter.

Thus, the main loops become:

```
for j in range(1, height-1):
    for i in range(1, width-1):
        # apply filter to pixel (i, j) in original image
```

Applying larger filters, such as 5×5, requires moving in two pixels from every edge.

EXERCISES

1. Write a function **apply(filter, img)** to return a copy of *img* with the *filter* applied. Assume *filter* is a function that returns an RGB tuple and takes two parameters: the original image *img* and a 2-tuple *pixel* of the pixel coordinates (i, j).

   ⟹ Caution: Do not modify the original image, because the filter depends on having the old values at every pixel.

2. Write a function **edge_filter(img, pixel)** to compute and return the RGB tuple for applying the edge detection filter to *img* at the 2-tuple $pixel = (i, j)$:

0	−1	0
−1	4	−1
0	−1	0

   Use it with the **apply()** function from the previous exercise to modify an image. Also try computing the negative of the result.

3. Write a function **blur_filter(img, pixel)** to compute and return the RGB tuple for applying the blur filter to *img* at the 2-tuple $pixel = (i, j)$:

1/8	1/8	1/8
1/8	0	1/8
1/8	1/8	1/8

   Remember that image RGB components must be integers.

4. Write a function **sharpen_filter(img, pixel)** to compute and return the RGB tuple for applying the sharpen filter to *img* at the 2-tuple $pixel = (i, j)$:

0	−1	0
−1	5	−1
0	−1	0

5. Write a function **emboss_filter(img, pixel)** to compute and return the RGB tuple for applying the emboss filter to *img* at the 2-tuple $pixel = (i, j)$:

2	0	0
0	−1	0
0	0	−1

6. Write a function `mean_filter(img, pixel)` to compute and return the RGB tuple for applying the mean or averaging filter to *img* at the 2-tuple *pixel* = (i, j):

1/9	1/9	1/9
1/9	1/9	1/9
1/9	1/9	1/9

CHAPTER 5

Objects and Classes

Python classes are powerful tools that allow us to define new types for creating customized objects. As a prelude to learning how to write classes, we review object-oriented terminology and how to use objects in programs.

5.1 USING TURTLE OBJECTS

To this point, you have quite a bit of experience working with Python objects, but many of these objects—especially strings, lists, and dictionaries—have specialized syntax because they are used so frequently. Turtle objects, as in Example 5.1, allow us to focus on common techniques for writing object-oriented code.

Code using objects is characterized by steps to create objects, call their methods, and occasionally access data attributes.

Object-Oriented Terminology

To review, most of these terms were introduced in Chapter 3:

Class A class is a template that defines objects of a new data type by specifying their state and behavior. For example, a `PlayingCard` class might define card objects for a game program.

Object An object is a specific **instance** of a class, with its own particular state. A program may create as many instances of a class as it needs. For example, a card game might create 52 `PlayingCard` objects to represent a deck of cards.

State An object's state is given by the current values of the data that it stores. Recall that the object state is stored in **data attributes**, also known as instance variables or fields. Objects are **immutable** if their state cannot change; otherwise, they

```
1  # ball.py
2
3  from turtle import Turtle, exitonclick
4
5  def main():
6      """Bouncing ball."""
7      ball = Turtle()
8      ball.shape("circle")
9      ball.penup()
10     ball.speed(0)
11     for _ in range(100):
12         ball.forward(5)
13         if ball.xcor() > 340:
14             ball.left(180)
15     exitonclick()
16
17 main()
```

Example 5.1 Bouncing ball.

are **mutable**. A PlayingCard would likely be an immutable object with attributes to store its rank and suit.

Behavior Object behaviors are specified by **methods**, sometimes known more specifically as **instance methods**. Methods are called **mutators** if they change the state of the object or **accessors** if they only return information about the state. For example, a CardDeck class might define a shuffle() mutator.

Constructor Constructors are called to create new objects. Each class has its own constructor of the same name as the class.

All data types in Python, including numeric types, are object types defined by classes.

Syntax: Creating Objects

The syntax to call a constructor and create a new object is:

```
ClassName(argument1, argument2, ...)
```

In Python, this is also known as **class instantiation**, since it creates a new instance of the class. The constructor's name is always the same as the name of the class, and new objects are usually named via assignment when they are created. A `Turtle` object is created and named `ball` in line 7 of Example 5.1.

Built-in types such as `list` and `str` often do not use constructors, because they have specialized literal syntax (`[1, 2, 3]` and `"abc"`) to create new objects. They do have constructors, though, listed in Table 5.1. These were described earlier as type converters, because that is often how they are used, but they are actually constructors for built-in classes.

⟶ Note: These built-in classes are the only class names that are lowercase. All other class names, such as `Turtle`, should be capitalized.

TABLE 5.1   Built-in constructors

`bool(x)` Create a boolean from x, if possible.
`dict(items)` Create a dictionary from *items*.
`float(x)` Create a float from x, if possible.
`int(x)` Create an integer from x, if possible.
`list(items)` Create a list from *items*.
`str(object)` Create a string from *object*.
`tuple(items)` Create a tuple from *items*.

As we move toward writing our own classes, calling constructors will be necessary because there will not be any specialized literal syntax for them.

The Turtle Class

The `turtle` module provides both procedural and object-oriented versions of its functionality. Chapter 1 used procedural code—meaning Python functions—because it is simpler to begin with. Example 5.1 introduces object-oriented `turtle` code using the `Turtle` class.

The `Turtle` class defines turtle objects that draw and move using method calls that are exactly the same as the function calls we used earlier. In fact, all functions in Tables 1.2, 1.3, and 1.5 except for `exitonclick()` are `Turtle` methods that can be called on `Turtle` objects. The one exception, `exitonclick()`, is a `Screen` method, but we will not need to create separate `Screen` objects.

In a sense, the methods come first, because the function calls are actually implemented in the background by calling the corresponding method on a single internal `Turtle` object. One advantage of using `Turtle` objects explicitly is that programs can then work with more than one turtle at a time.

Classes Hide Implementation Details

Using objects in a program allows you to think at the level of those objects while ignoring the details of how those objects actually work. Consider the `Turtle` object in Example 5.1. We do not know how its `forward()` method is implemented, but we have an idea of what to expect when it is called, and so we can think in terms of the turtle moving forward.

This is another form of **abstraction**, like writing functions (see page 15), where details are hidden to allow thinking at a higher level.

EXERCISES

1. Suppose a `Die` class has a constructor with no parameters, a `value` attribute and a `roll()` method. Write (on paper) code to create two dice, roll them, and print the sum of their values.

2. Suppose a `Die` class has a constructor with one parameter *sides* for its number of sides (the default is 6), a `value` attribute and a `roll()` method. Write (on paper) code to create two twenty-sided dice, roll them, and print the sum of their values.

3. Suppose a `Die` class has a constructor with no parameters, a `value` attribute and a `roll()` method. Write (on paper) code to create a list of five dice named `dice` and then roll all of the dice.

4. Suppose a `Card` class has a constructor with two parameters, an integer *rank* (2–14) and string *suit* (such as `"clubs"`). Write (on paper) code to create a list of all 52 cards named `deck` and then shuffle the deck.

5. Suppose an `RGB` class has a constructor with three integer parameters, *r*, *g*, and *b*, as well as a `luminance()` method that returns the color's relative luminance (see page 152). Write (on paper) code to create a color `darkblue` and print its luminance.

6. Suppose an `RGB` class has a constructor with three integer parameters, *r*, *g*, and *b*, as well as a `luminance()` method that returns the color's relative luminance (see page 152). Write (on paper) code to create two different colors (your choice) and print the sum of their luminance.

7. Modify Example 5.1 to have the ball move farther and bounce off both sides of the screen. Do not worry about touching the sides of the screen precisely.

8. Modify Example 5.1 to add a second turtle that moves more slowly than the ball and has the "turtle" shape. Have both the ball and turtle bounce off both sides of the screen, with the ball higher on the screen than the turtle. Do not worry about touching the sides of the screen precisely.

9. Modify Example 5.1 to have the ball move up and down instead of side-to-side, bouncing off both the top and bottom of the screen. Do not worry about touching the edges of the screen precisely.

10. Modify Example 5.1 to have the ball start moving at a random angle and bounce off all four sides. Do not worry about touching the edges of the screen precisely.

11. Modify Example 5.1 to give the ball a random size and color. The `color()` method uses float RGB values in $[0, 1]$ rather than integers. Do not worry about taking the size into account for bouncing.

12. Write an object-oriented turtle graphics program to draw something of your choice.

5.2 WRITING CLASSES

Although Python comes with a rich set of built-in types, no programming language can anticipate all data types that a programmer might eventually want, and so object-oriented languages allow you to create new data types by defining classes. Example 5.3 shows how to create a new type to represent six-sided dice.

Syntax: Class Definitions

Classes are used to define new types in order to help manage the complexity that comes with building larger, more realistic programs. All class definitions begin with one line:

```
class ClassName:
    body
```

Class names are capitalized, using the "CapitalizeEachWord" convention. This helps distinguish class names from ordinary variable and function names, which are lowercase. The *body* of a class definition consists of a sequence of method definitions. In Example 5.2, the body of the `Die` class contains three method definitions.

```
1   # rolldice.py
2
3   from random import randint
4
5   class Die:
6       """A single 6-sided die."""
7
8       def __init__(self):
9           """Initialize to a random value."""
10          self.value = randint(1, 6)
11
12      def roll(self):
13          """Simulate rolling the die."""
14          self.value = randint(1, 6)
15
16      def __str__(self):
17          return str(self.value)
18
19  def main():
20      die1 = Die()
21      die2 = Die()
22      for _ in range(10):
23          die1.roll()
24          die2.roll()
25          print(die1, die2)
26
27  main()
```

Example 5.2 Rolling dice.

Defining Methods

Methods are defined inside the body of a class in the same way as functions, using **def**, except that the first parameter for every method should be a variable named **self**. In other words, only two things distinguish an instance method from a regular function: the **def** appears inside the body of a class, and the first parameter is **self**. Methods are always called on an object, so think of the **self** parameter as the object on which the method is called.

Defining Data Attributes

Python does not provide a way to designate ahead of time that a variable is a data attribute of an object. Instead, inside the methods of a class definition, any variable name prefixed with "`self.`" is considered to be a data attribute and is stored within the object.

Values are given to attributes with normal assignment statements:

```
self.attribute = value
```

Think of the attribute as belonging to the instance, keeping in mind that the name of the instance inside of method definitions is `self`. Although new attributes may be created at any time, it is a good idea to give all data attributes a valid starting value in the `__init__()` method (see below).

The `Die` class of Example 5.2 has one data attribute named `value`. Every `Die` object has its own separate storage location for its attributes, so different dice each store a different `value`, as shown in Fig. 5.1.

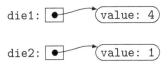

Figure 5.1 Data attributes.

Special Methods: `__init__()`

Python provides class customization with **special methods** whose names start and end with two underscores (`__`). These methods are called automatically in different situations, depending on their name. The `__init__()` method is called whenever the constructor is called to create a new object. The purpose of `__init__()` is to **initialize** the object by giving appropriate starting values to each of its data attributes. Any parameters sent to the constructor are passed on to the class's `__init__()` method.

Every class definition should include an `__init__()` method. In Example 5.2, the `__init__()` method of the `Die` class gives an initial value to the `value` attribute of each new `Die` object.

Special Methods: `__str__()`

The `__str__()` method is called any time the object appears in a `print()` statement or as an argument to `str()`. The `__str__()` method should return a string that describes the object. For example, the `Die` class in Example 5.2 returns a string containing the current value of the die. A `__str__()` method is not required, but it is helpful for many classes.

Method Calls inside a Class

If you want to call an instance method from within a different method in the same class, the syntax is similar to accessing data attributes:

```
self.method(arguments)
```

Technically, methods are also attributes of the object, and so that is why they are accessed in the same way as data attributes. The only difference is the use of parentheses to indicate the method call.

For example, you may have noticed that line 10 of the __init__() method in Example 5.2 is the same as line 14 of the roll() method. That means we could rewrite __init__() to call self.roll() instead of setting self.value directly:

```
def __init__(self):
    """Initialize to a random value."""
    self.roll()
```

The original version was chosen to reinforce the importance of initializing data attributes in __init__(), but this version does that equally well.

⟶ Note: It is easy to forget "self." in these method calls. Without it, Python assumes you are calling a regular function instead of an instance method.

Writing inside vs. outside the Class

Pay attention to the differences between code inside the Die class of Example 5.2 and code that uses the Die class in main():

> **Inside** the Die class, instance methods all have a self parameter and access attributes and other methods using "self."
>
> **Outside** the Die class, the main() function *uses* the class, as in Section 5.1 and all prior chapters.

In particular, there are no references to self outside the class.

EXERCISES

1. Modify Example 5.2 to roll three dice instead of two. Ask the user for the number of rolls.

2. Modify Example 5.2 to roll n dice instead of two. Ask the user for the number of dice n and the number of times to roll them.

3. Modify the Die class in Example 5.2 to allow initializing dice with different numbers of sides, with a default value of 6. Include a program to test your class, asking the user for the number of sides on each of two dice, as well as the number of times to roll them.

4. Modify Example 5.2 to add the values of the pair of dice and keep track of how many times each sum is rolled. Ask the user for the number of rolls, and report the count for each sum at the end.

5. Write a Coin class to represent a coin that can be flipped. Store whether heads or tails is face up with an integer that is 0 or 1. Write a flip() method to flip the coin, an is_heads() method to return True if the coin is currently showing heads (otherwise, False), and a __str__() method that returns "H" or "T". Include a program to count the number of heads in a given number of flips. Ask the user for the number of times to flip the coin.

6. (Requires Exercise 5.) Use the Coin class from the previous exercise to write a program that counts the number of coin flips necessary to reach a given number of heads in a row. Ask the user for how many consecutive heads to wait for, and report the total number of flips needed at the end.

7. Write a Circle class to represent a circle, initialized with its radius. Store the radius, and include an area() method to return the area of the circle, a circumference() method to return the circumference, and a __str__() method. Write a main() function that creates a few circles and prints each area and circumference.

8. Write an RGB class to represent an RGB color, initialized with its red, green, and blue components. Include a __str__() method and a luminance() method that returns the luminance of the color. Write a program to test your class.

9. Write a Point2D class to represent a point (x, y) in two dimensions. Include a __str__() method and a distance_to(other) method that returns the distance to another point. Write a program to test your class.

10. Write a Vector2D class to represent a two-dimensional vector (x, y). Include a __str__() method, as well as a norm() method that returns the length of the vector, $\sqrt{x^2 + y^2}$. Write a main() function to test your code.

11. Write a Date class to represent a date initialized with integers for day, month, and year. Include a __str__() method that returns an appropriate string format and an is_before(other) method that returns True if this date comes chronologically before the other date. Write a main() function to test your class.

12. Write an Account class to represent a bank account initialized with an account number and starting balance. Include deposit(amount) and withdraw(amount) methods. Withdrawals should only be performed if the balance is high enough. Include the account number and balance in the __str__() method, and include a program to test your class.

13. Write an `Item` class to represent one item in an inventory. Store the name, quantity, and price of the item, and provide an `add(number)` method to add to the quantity in stock and a `sell(number)` to sell items, as long as there is sufficient quantity in stock. Also write a `value()` method that returns the total value of the current stock of this item. Include the name, quantity, and total value in the `__str__()` method, and include a program to test your class.

14. Write a `Card` class to represent an immutable playing card from a standard 52-card deck, initialized with its numeric rank (2–14) and a suit that is hearts, diamonds, spades, or clubs. Include a `__str()__` method that returns a shorthand string representation of the card, such as `"5C"` for the five of clubs or `"JD"` for the jack of diamonds. Write a `main()` to create a list of all 52 cards, shuffle the list, and then print the cards in shuffled order. Use a loop over the list of cards to print them.

5.3 COMPOSITION

Classes are rarely used in isolation. Their power comes from interactions between objects, and determining how objects will work together is the art of object-oriented design. Example 5.3 shows one way to build a `Dice` class to represent a group of dice from the `Die` class in the previous section. Comments are omitted to fit on the page.

In a strong sense, nothing new is required for the `Dice` class to use the `Die` class: once `Die` has been written, it can be used like any other type.

Class Relationships

Two of the basic relationships between classes in object-oriented languages are known as has-a and is-a:

> **Has-a** relationships are when an object of one type *has an* object (or collection of objects) of another type as one of its data elements.
>
> **Is-a** relationships are when a type *is a* specialized version of another type.

For example, each `Dice` object of Example 5.3 *has a* list of `Die` objects stored in its data attribute `self.dice`. It would not make sense to say that a `Dice` object "is a" `Die`. Similarly, each `Die` object *has an* integer `self.value` data element.

Python has a `Number` type to represent all kinds of numeric values, so an example of an is-a relationship is that an `int` *is a* `Number`, and likewise, a `float` *is a* `Number`. Neither of these fits a "has-a" relationship.

```python
1  # dice.py
2
3  from random import randint
4
5  class Dice:
6      def __init__(self, n):
7          self.dice = [Die() for _ in range(n)]
8
9      def rollall(self):
10          for die in self.dice:
11              die.roll()
12
13      def values(self):
14          return [int(die) for die in self.dice]
15
16      def __str__(self):
17          return str(self.values())
18
19  class Die:
20      def __init__(self):
21          self.value = randint(1, 6)
22
23      def roll(self):
24          self.value = randint(1, 6)
25
26      def __str__(self):
27          return str(self.value)
28
29      def __int__(self):
30          return self.value
31
32  def main():
33      dice = Dice(5)
34      dice.rollall()
35      print(dice)
36
37  if __name__ == "__main__":
38      main()
```

Example 5.3 Dice class.

Has-a relationships are modeled by composition, discussed next, whereas is-a relationships are modeled by inheritance, which is the subject of Section 5.5.

Composition Models Has-A Relationships

Has-a relationships are modeled by **composition**, which is just storing an object (or collection of objects) of the other type in an instance variable. This should seem natural, since if object A "has an" object B, then B is stored somewhere as an attribute of A. As noted above, each `Dice` object has a list of `Die` objects stored in its `self.dice` attribute, and each `Die` has an integer stored in its `self.value`.

This type of modeling is also sometimes known as aggregation.

Special Methods: __int__()

The __int__() special method is called any time an object appears as an argument to the `int()` function. If implemented, it should return an integer that meaningfully represents the object. The `Die` class implements __int__() so that `Die` objects can be converted to their integer values—that is, the number showing on the die stored in `self.value`.

The `Dice` class does not implement __int__(), since no one integer usually represents a group of dice. Instead, a list of the values on the set of dice is more natural.

Converting Dice to a List of Values

In Example 5.3, the `Dice` class stores a list of `Die` objects in `self.dice`, but there are situations where it would be helpful to have a list of the integer values of the dice instead of the `Die` objects. The `values()` method of the `Dice` class is designed for this purpose. It works by calling `int()` on each of the individual `Die` objects in a list comprehension on line 14. (There is also another list comprehension in line 7.) See pages 76 and 85 to review list comprehensions.

The `values()` method makes it easy to write __str__() for the `Dice` class, because it can just convert the list of integers to a string. Notice the similarity this provides to the __str__() method of the `Die` class.

⟶ Note: In other languages, a `value()` method might also be written in the `Die` class to return the value, but for a class like this in Python, it is more common to just access the `value` attribute.

Special Attributes: __name__

One aspect of Example 5.3 is in preparation for its use in the next section. The **if** statement on line 37 is new: it essentially tests to see if this program is being run itself, or if it is being imported by another program.

The special variable __name__ is set to the string "__main__" when this file is run as a program itself. Otherwise, if this file is imported by another program, __name__ will be set to "dice", the name of this file (without the ".py"). Thus, the **if** statement on line 37 only runs the **main()** function when this file is run as the "main" program. If this file is imported by another program, then this file's **main()** will not run.

Thus, it is common for Python classes that may be imported to have a **main()** that tests the class, along with an **if** statement to make sure the tests do not run when the file is imported.

EXERCISES

1. Suppose you plan to design **Student**, **Course**, and **Instructor** classes for a university registration system. List some states and behaviors that might be included in each class. Then list any has-a or is-a relationships that exist between these classes.

2. Suppose you plan to design **Employee**, **Department**, and **Manager** classes for a human resources system. List some states and behaviors that might be included in each class. Then list any has-a or is-a relationships that exist between these classes.

3. Rewrite the **Dice** class initializer in Example 5.3 without using list comprehension. Discuss the tradeoffs.

4. Rewrite the **Dice** class **values()** method in Example 5.3 without using list comprehension. Discuss the tradeoffs.

5. Modify Example 5.3 to add a **roll(i)** method to the **Dice** class to roll only the *i*-th die. Test your method.

6. Modify Example 5.3 to add a **total()** method to the **Dice** class to return the sum of the values of the dice. Test your method.

7. (Requires Exercise 6.) Using the **total()** method of the previous exercise, write a program to simulate rolling two dice, keeping track of the number of times each total occurs. Ask the user for the number of rolls, and report the number of times each total occurs at the end.

8. Repeat the previous exercise with *n* dice, asking the user for the value of *n*.

9. Use the `Coin` class from Exercise 5 in Section 5.2 to write a `Coins` class that represents a group of *n* coins. Include a `flipall()` method to flip all the coins, a `heads()` method to return the number of coins that currently show heads, and a `__str__()` method. Write a program to test your class.

10. (Requires Exercise 9.) Use the previous exercise to write a program to simulate flipping *n* coins, keeping track of the number of heads that occur each time. Ask the user for the number of coins to use and the number of times to flip them, and report the results at the end.

11. Use the `Point2D` class from Exercise 9 in Section 5.2 to write a `Segment` class that represents a line segment between two points. Include a `length()` method that returns the length of the segment and a `__str__()` method, as well as a program to test your class.

12. Use the `Point2D` and `Vector2D` classes from Exercises 9 and 10 in Section 5.2 to write a `Ray2D` class. A **ray** has a `Point2D` origin *o* and a `Vector2D` direction $\vec{d}$. Write a `point_at(t)` method to return the `Point2d` on the ray at a given *t*:

$$o + t \cdot \vec{d}$$

Include a `__str__()` method and a program to test your class.

13. Use the `Item` class from Exercise 13 in Section 5.2 to write an `Inventory` class representing a collection of items in inventory. Store the items in a dictionary keyed by the item name. Write a `__str__()` method, along with these two methods:

`inv.add(item)`
Add `Item` *item* to inventory if new; otherwise, increase the quantity on hand by the number in *item*.

`inv.sell(itemname, quantity)`
Reduce number of items named *itemname* by given *quantity*, as long as there is sufficient stock on hand.

Include a program to test your class.

14. Use the `Card` class from Exercise 14 in Section 5.2 to write a `Deck` class representing a deck of 52 playing cards. Include a `shuffle()` method and a `__str__()` method. Test your code. Hint: for the `__str__()` method, use a list comprehension that applies `str()` to each card.

5.4 IMPORTING CLASSES

As programs become more complicated and use more types, it can be helpful to isolate different families of code into separate files. Example 5.4 shows how the `Dice` class from the previous section can be used to create a simple game written in another file.

Chuck-a-Luck is played with three dice. The player guesses a number between 1 and 6, and the dice are rolled. If the number appears (either 1, 2, or 3 times), then the player wins that many points; otherwise, the player loses one point.

To run this program, put both `chuck.py` and `dice.py` from the previous section in the same folder. In `chuck.py`, the name "Dice" is imported "**from dice**." That means this program will look for the definition of `Dice` in a file named `dice.py`. The **import** statement is the link between the two files.

Importing from Your Own Modules

When you write a class like `Die` or `Dice`, it would be nice to be able to write it once and then use it anytime you need to represent dice in a game. This is called **code reuse**. The way to reuse code in Python is to write the class definitions or functions in a separate file, called a **module**, and then **import** the definitions you need from that module into whatever program wants to use them.

The syntax is the same as it is to import a library module:

```
from module import name1, name2, ...
```

There is nothing special about Python module files, except they have definitions you would like to use in other files. They may contain more than one class definition (as in Example 5.3), as well as function definitions.

The name of a module file must end with ".py," but do not include ".py" in the name of the module in the **import** statement. Put the module file in the same folder as the program that wants to use it.

Syntax: Conditional Expressions

Fairly often, you will find yourself wanting to set a variable or return a value based on a boolean condition. For example, you may write this:

```
if condition:
    variable = value1
else:
    variable = value2
```

```
 1  # chuck.py
 2
 3  from dice import Dice
 4
 5  class ChuckALuck:
 6      """Play the game Chuck-a-Luck."""
 7      def __init__(self):
 8          self.dice = Dice(3)
 9          self.score = 0
10
11      def play_once(self, guess):
12          """Roll dice and update score based on guess."""
13          self.dice.rollall()
14          print("The roll:", self.dice)
15          matches = self.dice.values().count(guess)
16          score = matches if matches > 0 else -1
17          print("Score:", score)
18          self.score += score
19
20      def play(self):
21          """Allow user to repeatedly guess."""
22          print("Welcome to Chuck-A-Luck")
23          guess = int(input("Guess 1-6 (0 to stop): "))
24          while guess != 0:
25              self.play_once(guess)
26              print("Current total:", self.score)
27              guess = int(input("Guess 1-6 (0 to stop): "))
28          print("Thanks for playing.")
29
30  def main():
31      game = ChuckALuck()
32      game.play()
33
34  if __name__ == "__main__":
35      main()
```

Example 5.4 Chuck-a-Luck.

Python offers a shorthand **conditional expression** syntax that can be easier to read:

```
value1 if condition else value2
```

The value of this expression is *value1* if the *condition* is True; otherwise, it is *value2*. It allows you to rewrite the above **if** statement like this:

```
variable = value1 if condition else value2
```

Notice the assignment that makes the conditional expression into a complete statement: it is the part that is common to both the **if** and **else**:

```
if condition:
    ⟨variable =⟩ value1
else:
    ⟨variable =⟩ value2
```

becomes:

```
⟨variable =⟩ value1 if condition else value2
```

Conditional expressions are also commonly used in **return** statements, as in the following example.

Example: mymin2(x, y)

The mymin2(x, y) function, which returns the smaller of two values, may be written with a conditional expression:

```
def mymin2(x, y):
    return x if x <= y else y
```

This version is more succinct, and in some ways more natural, than the equivalent **if-else** statement.

Abstraction

Look again at Example 5.4. By hiding the details of how dice work within the Dice class, we can focus on understanding the game itself. This is another example of abstraction, since we ignore the details of how the dice actually work. We are able to work at a higher level of abstraction than if the code included all of those details in the middle of it. One of the tasks in object-oriented design is to create well-defined classes to represent important objects in the application domain.

EXERCISES

1. Rewrite using a conditional expression:

```
if condition:
    return value1
else:
    return value2
```

2. Rewrite line 16 of Example 5.4 without the conditional expression.

3. Explain line 15 of Example 5.4. As part of your answer, draw a diagram like the one on page 125 for combining method calls.

4. Identify the relationship between the ChuckALuck and Dice classes: is it has-a, is-a, or neither? Explain your answer.

5. Write a myabs(x) function to return the absolute value of x using a conditional expression. Include a main() to test your function.

6. Rewrite the twotone(img, dark, light) function of Exercise 12 in Section 4.2 using a conditional expression to choose between dark and light at each pixel.

7. Write a program to play a one-player version of the dice game Pig. The player rolls one die until either a 1 is rolled or the player decides to stand. If a 1 is rolled, the player scores 0 for that turn; otherwise, when the player stands, he or she scores the sum of the rolls during that turn. The game continues until the player reaches a score of 100.

8. Write a program to simulate shuffling and dealing a complete deck of cards to two players. Add these methods to the Deck class from Exercise 14 of Section 5.3: dealcard(player) to remove the top card from the deck and give it to player, and empty(), which returns True if the deck is empty. Write a Player class with a receivecard(card) method to add card to the player's hand, and a __str__() method to see the hand. Print the contents of the deck (which should be empty) and both players' hands after the deal.

9. Write a program to play a simplified version of the dice game craps. In each round, two dice are rolled. The player wins on 7 or 11; the player loses on 2, 3, or 12. Any other total establishes a *point*, where the dice continue to be rolled until either the player wins by rolling the point again, or loses by rolling 7. Show each roll of the dice and the player's score at the end of each round.

10. Write a program to play a simplified version of dice poker. In each round, five dice are rolled. Scoring hands are five of a kind, a straight, four of a kind, a full house, three of a kind, or two pair. Give each of these an appropriate value, and report the player's running score at the end of each round.

11. Modify the previous exercise to allow the player one chance in each round to choose some or all of the dice to re-roll (including none).

5.5 INHERITANCE

Classes may also be related to each other through a mechanism called inheritance. When one class inherits from another, it inherits all the state and behavior of the original class, and then has the opportunity to define new additional state and behavior.

Example 5.5 rewrites Example 5.1 using inheritance to create a specialized kind of turtle, a bouncing ball.

Inheritance Models Is-A Relationships

Recall from Section 5.3 that is-a relationships are when one type *is a* specialized version of another type. The `Ball` class of Example 5.5 is a specialized turtle: it can do everything a turtle can do, but it has a specific shape and speed, as well as a new `move()` method. Is-a relationships are modeled in object-oriented languages with inheritance.

Inheritance

When we define a class using **inheritance**, the new class inherits all of the state and behavior of the original class, known as the **base class**. The new class is called an **extension** of the base class, because it may extend the base class with additional new state and behavior. Base classes are also known as **parent** classes or **superclasses**; extensions are also known as **child** classes or **subclasses**.

The syntax to define a class as an extension of a base class in Python is to put the name of the base class inside parentheses:

```
class ClassName(BaseClass):
    body
```

The body in this definition may be empty, in which case the extension is identical to the base class.

```
1  # ballclass.py
2
3  from turtle import Turtle, exitonclick
4
5  class Ball(Turtle):
6      """Round turtle that bounces."""
7      def __init__(self):
8          super().__init__()
9          self.shape("circle")
10         self.penup()
11         self.speed(0)
12
13     def move(self):
14         """Move forward, bouncing against right side."""
15         self.forward(5)
16         if self.xcor() > 340:
17             self.left(180)
18
19  def main():
20      ball = Ball()
21      for _ in range(100):
22          ball.move()
23      exitonclick()
24
25  if __name__ == "__main__":
26      main()
```

Example 5.5 Bouncing ball class.

In Example 5.5, the Ball class extends the Turtle base class. Because of this, all of the methods defined in the Turtle class are available to be called on any Ball object. In this way, a Ball object "is a" Turtle.

Defining New Methods or State

When using inheritance, the subclass can define as many new methods as it needs. Similarly, it may define new data attributes, simply by assigning them values, as usual. These new methods and attributes will only be available on instances of the subclass; the superclass has no knowledge of them. In Example 5.5, the Ball class defines a new move() method, but adds no new state.

Overriding Methods

Class extensions are also allowed to redefine existing methods from the base class, known as **overriding** the method from the base class. Most often, this involves adding new behavior to the existing method, so there needs to be a way—inside the definition of the new class—to call the old method that belongs to the superclass. That is the role of the built-in **super**() function, listed in Table 5.2.

TABLE 5.2 The **super**() function

super()
Return an object that gives access to the superclass type.

To use **super**(), just call superclass methods on the object it returns:

```
super().method(arguments)
```

⟶ Note: The call **super**() takes the place of **self** in internal method calls, so **self** is *not* used as the first argument of these method calls.

Overriding __init__()

The **Ball** class in Example 5.5 overrides the __init__() method from its superclass, **Turtle**. Any time you override __init__(), it is good practice to allow the superclass to initialize the object first. Thus, there is a call to **super**().__init__() on line 8.

Tuples in Turtle Methods

Some **Turtle** methods are flexible in that they can take either separate x and y coordinate parameters or tuple (x, y) parameters. The goto(), towards(), and distance() methods all work this way. For example, these two calls of towards() are equivalent for a **Turtle** object **t**:

```
t.towards(10, 45)
point = (10, 45)
t.towards(point)
```

In addition, the towards() and distance() methods may be given **Turtle** parameters, in which case, they return the heading and distance to the other turtle, respectively.

Drawing Faster Turtles

For complex drawings, turtle graphics can be slow because they animate each step of each turtle on the screen. The tracer() function, listed in Table 5.3, can help speed up drawing by reducing the number of times the screen is updated. For example, with 100 turtle objects on the screen, a call to

```
tracer(100, 0)
```

will draw the screen once for all 100 turtles with no delay, instead of updating the screen for each turtle.

TABLE 5.3   `turtle` module: tracer

`tracer(n, delay)`
Draw turtles, updating screen only every *n* frames with given *delay* in milliseconds.

EXERCISES

1. Finish the `Ball` class in Example 5.5 so that it bounces off all four sides when it moves. Do not worry about hitting the sides exactly. Test your class.

2. Using the `Ball` class from the previous exercise, modify `main()` to create and move 3 different balls. Start each in a different direction.

3. Modify the `Ball` class to add a `pace` attribute that specifies the distance the ball travels forward each time it moves. Set the starting value of `pace` to 5. Test your class.

4. Extend the `Ball` class to create a `RandomBall` subclass with random size, color, heading, and pace (see the previous exercise). The `Turtle color()` method uses float RGB values in $[0, 1]$ instead of integer RGB. Write a `main()` to create and bounce a list of 100 random balls.

5. Write a `SquareTurtle` class extending `Turtle` that has a `draw(width)` method to draw a square of the given *width* starting at the turtle's current position. Include a program to test your class.

6. Write a `PolyTurtle` class extending `Turtle` that has a `draw(sidelength)` method to draw a polygon of the given *sidelength* starting at the turtle's current position. The `PolyTurtle` constructor takes an integer parameter giving the number of sides in the polygon. Include a program to test your class.

7. Write a `WrappingTurtle` class extending `Turtle` with a `move()` method that moves the turtle on screen, wrapping across to the other side of the screen when reaching an edge. Include a `pace` attribute specifying how far the turtle travels each time it moves. Write a `main()` to create and move several wrapping turtles.

8. Design a `WanderingTurtle` class extending `Turtle` with a `move()` method so that repeatedly moving the turtle causes it to wander randomly around the screen. Include a program to demonstrate your class.

9. Write a `TrackingTurtle` class extending `Turtle` with a `move()` method that will cause the turtle to move toward another turtle it is tracking. The turtle to be tracked should be a parameter to the constructor. Test your turtle on a wandering turtle from the previous exercise.

10. Write a `ThreePointTurtle` class extending `Turtle` whose constructor takes three points in the plane, *A*, *B*, and *C*. It then provides a `draw(n)` method implementing the following algorithm:

    Draw dots at A, B, and C
    Repeat n times:
      Choose one of A, B, or C and move halfway toward it
      Draw a dot

    Include a `halfway_to(point)` helper method to move the turtle halfway to *point*. Test your class, first with small values of *n*.

11. Design your own extension of the `Turtle` class. Include a program to demonstrate its use.

5.6   GRAPHICAL USER INTERFACES

Most of the programs we have written so far use a **text-based interface**, meaning that user input and output are via a screen that only displays text. Turtle graphics programs have been the only exceptions. However, most of the applications you use regularly are written with a **graphical user interface (GUI)** instead. Graphical interfaces occupy their own windows and can contain complex layouts of buttons, sliders, text boxes, and other components, as well as menus such as File, Edit, and Help.

Example 5.6 uses the `tkinter` module to build a simple temperature conversion program. Because GUI applications require more code than comparable text-based programs, its listing is somewhat compressed to fit on the page.

$\longrightarrow$ Note: The GUI window may not close after clicking the Quit button if run from IDLE, since IDLE is itself a `tkinter` application. In that case, close the window manually after clicking Quit.

The `tkinter` Module

The `tkinter` module is one of the standard libraries for developing graphical applications in Python. It provides an interface to Tk, which is a cross-platform

```python
from tkinter import *

class TempConverter(Frame):
    def __init__(self, parent):
        super().__init__(parent)
        self.parent = parent
        parent.title("Temperature Converter")
        self.grid(row=0, column=0, padx=80, pady=60)
        self.fahr = DoubleVar()
        self.fahr.set(32)
        self.cels = StringVar()
        self.refresh(0)
        self.create_widgets()

    def create_widgets(self):
        self.cels_lbl = Label(self, textvariable=self.cels)
        self.cels_lbl.grid(row=0, column=0, pady=30,
                           columnspan=2)
        self.f_lbl = Label(self, text="deg F:")
        self.f_lbl.grid(row=1, column=0, pady=10, sticky=E)
        self.f_scl = Scale(self, variable=self.fahr, to=500,
                           from_=-100, command=self.refresh,
                           length=400, orient=HORIZONTAL)
        self.f_scl.grid(row=1, column=1, padx=10, pady=10)
        self.quit = Button(self, text="Quit", padx=20,
                           pady=5, command=self.parent.quit)
        self.quit.grid(row=2, column=1, sticky=E, padx=10,
                       pady=30)

    def refresh(self, _):
        f = self.fahr.get()
        c = (5/9)*(f - 32)
        self.cels.set("{:.2f} deg C".format(c))

root = Tk()
gui = TempConverter(root)
root.mainloop()
```

Example 5.6 Temperature converter GUI.

GUI library for the Tcl language. Because a typical GUI application uses many components from `tkinter`, the *-form of the `import` is common.

This section contains only a brief introduction to `tkinter`, so consult the documentation for more information.

The Tk Root and Event Loop

The `Tk` class from the `tkinter` module creates the main window for a GUI application. Once a `Tk` object has been created, then GUI components called widgets are added to it, and the `Tk` object's `mainloop()` method is called.

The `mainloop()` method starts an **event loop**, during which the application listens for and responds to user events, such as mouse clicks, clicked buttons, and selecting menu items. Generally, event loops run until the `quit()` method is called. Table 5.4 lists these methods for controlling the event loop in `tkinter`, along with some of the other methods common to all widgets.

TABLE 5.4 `tkinter` module: widget methods

`w.config(options)`
Configure widget with new *options*.
`w.grid(options)`
Put widget on screen using grid manager and specified *options*.
`w.mainloop()`
Start event loop.
`w.quit()`
Stop event loop.

The event loop is also stopped if the user closes the main window of the application. Programming based on an event loop, responding to user events in this way, is known as **event-driven programming**.

The root `Tk` object sets the title of the application window by calling its `title()` method, as in line 7 of Example 5.6. This method, listed in Table 5.5, may be called on any top-level window.

TABLE 5.5 `tkinter` module: `Toplevel` window method

`w.title(text)`
Set window title to *text*.

Extending Frame and the Component Hierarchy

Once a `Tk` root is created, it is common to subclass the `Frame` class to build the application. `Frame` widgets are designed to group components on screen.

They do this by adding other widgets as **children**, creating a hierarchy of all the components in the application. In Example 5.6, the `TempConverter`, as a `Frame`, is a child of the root, and then the other components are added as children of the `TempConverter`, as shown in Fig. 5.2.

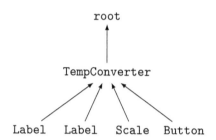

Figure 5.2 GUI component hierarchy.

These parent-child relationships are established when widgets are created, because the first argument to a widget's constructor is always its parent. Thus, `root` becomes the `TempConverter`'s parent by being passed as an argument to the constructor in line 36 and then to the superclass `Frame`'s `__init__()` in line 5. All of the other widgets become children of the `TempConverter` by having `self` as the first argument to their constructor calls, such as on line 16 for the first `Label`.

⟹ Caution: Assigning `self.parent` to `root` on line 6 of the `__init__()` method does *not* make the root this object's parent. It only saves the root in an instance variable so that the `quit()` method can be called on it later in line 26.

Widget Classes

GUI components in `tkinter` are known as **widgets**. The widget classes used in Example 5.6 are listed in Table 5.6. Many other widget classes are described in the full `tkinter` documentation.

Again, the first argument to all widget constructors is its parent component. Because these constructors have so many other possible parameters, other arguments are passed as **keyword** arguments that take this form:

```
keyword=value
```

The advantage of using keywords is that the caller does not need to list every argument in the precisely correct order. Instead, the caller may choose to send only some arguments, and they may be listed in any order. This is ideal for setting the large number of possible widget options.

TABLE 5.6 `tkinter` module: widget classes

`Button` Button to click.
`Entry` Allow user to type one line of text.
`Frame` Container to hold other widgets.
`Label` Display text or image.
`Scale` Slider to set a numeric value.

TABLE 5.7 `tkinter` module: `Label` options

`text` String to display on widget.
`textvariable` `StringVar` control variable (see below) to display on label.

Widget Options

Different options are available for each type of widget, although a few are common to all widgets. Tables 5.7, 5.8, and 5.9 describe the options used in Example 5.6 (without repeats).

Options are set using keyword arguments when the widget is created or may be updated later by calling the widget `config()` method.

⟶ Note: The `command` option of the `Scale` widget sends the current value of the slider as a string parameter to the specified command, which is a strange choice. This string is ignored by our `update()`, and the control variable is used to get the value of the slider instead.

TABLE 5.8 `tkinter` module: `Button` Options

`command` Function or method to call on click, with no parameters.
`padx` Extra horizontal space around contents.
`pady` Extra vertical space around contents.

TABLE 5.9   `tkinter` module: `Scale` options

`command`
Function or method to call when slider moves, with one string parameter.
`from_`
Left or top value of slider. (Underscore distinguishes from the keyword.)
`length`
Length of slider in pixels.
`orient`
`VERTICAL` (default) or `HORIZONTAL`.
`to`
Right or bottom value of slider.
`variable`
Control variable linked to value of slider.

Control Variables

The `tkinter` module provides **control variables** to link between widgets and the rest of the GUI application. They come in three types, listed in Table 5.10.

TABLE 5.10   `tkinter` module: control variable types

`DoubleVar`
Floating-point variable that can be linked to a widget.
`IntVar`
Integer variable that can be linked to a widget.
`StringVar`
String variable that can be linked to a widget.

Control variables are linked to widgets by options such as `variable` for `Scale` and `textvariable` for `Label`. When a control variable is linked to a widget, then changes in the variable update the widget automatically, just as changes by the user to the widget update the variable. Applications then access control variables using the `get()` and `set()` methods from Table 5.11. See Fig. 5.3.

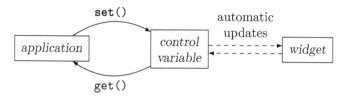

Figure 5.3   Control variable communication.

⟹ Caution: It is easy to forget to call `get()` and `set()` on control variables.

TABLE 5.11   `tkinter` module: control variable methods

v.`get`()
Return value of variable.
v.`set`(*newvalue*)
Set value of variable to *newvalue*.

Grid Layout

All GUI applications need a way to define how their components should be laid out in the application window. Example 5.6 uses the **Grid manager** to place widgets into rows and columns. Each widget calls its `grid()` method with options from Table 5.12 to specify its position. Row-column cells are numbered starting at 0, and each widget normally occupies one cell.

⟹ Caution: Widgets will not appear until their `grid()` method is called.

Rows and columns will be sized large enough to fit each widget in its cell. The `sticky` option controls what happens to extra space in the cell beyond the widget's natural size. If no `sticky` option is given, the widget is centered in its cell's available space. Otherwise, the specified directions pin the widget's edges to those edges of the cell.

TABLE 5.12   `tkinter` module: grid manager options

`column`
Column number to place widget in.
`columnspan`
Number of columns to occupy (if more than one).
`padx`
Extra horizontal space around widget.
`pady`
Extra vertical space around widget.
`row`
Row number to place widget in.
`rowspan`
Number of rows to occupy (if more than one).
`sticky`
Edges of the widget to pin to boundary of available space.

Example: `sticky=EW`

Pins both the left and right edges of the widget to the left and right edges of the cell, causing the widget to stretch horizontally (if necessary) to fill the cell's available width.

EXERCISES

1. Describe the differences between using the padding options of a widget and padding options in the `grid()` method.

2. Draw a picture of the layout for Example 5.6, showing the rows and columns, padding, and stickiness.

3. Modify Example 5.6 to convert temperatures from Celsius to Fahrenheit. Rename variables appropriately.

4. Write a GUI application to convert positive decimal integers on a slider to binary and hexadecimal. Display the binary and hex values in labels above the slider, and provide a Quit button.

5. Write a GUI application for the `complement()` function of Example 3.2 in Section 3.2. Use an `Entry` widget for the user to enter the strand of DNA, and provide buttons to Generate (and display) the complement, and Quit.

6. Write a GUI application for the `conjugate()` function of Example 3.1 in Section 3.1. Use an `Entry` widget for the user to enter the verb, and provide buttons to Generate (and display) the conjugation, and Quit.

7. Write a GUI application for the `username()` function of Exercise 18 in Section 3.1. Use `Entry` widgets for the user to enter the first and last names, and provide buttons to Generate (and display) the username, and Quit.

8. Write a GUI application for the `years_to_goal()` function of Example 2.4. Provide sliders for the principal, interest rate, and goal, and update the number of years to reach the goal if any of the sliders change. Include a button to quit the application.

9. Write a GUI application to allow the user to guess a random number between 1 and 100. Use an `Entry` widget for the user to enter the guess, and provide buttons to Reset the game, Check the guess (indicating if it is too high, too low, or correct), and Quit. Display the number of guesses after a correct response.

10. Write a GUI application to display RGB colors, where the red, green, and blue values are set by sliders. Display the color as the background color of a `Label` with fixed width and height. Colors are specified as hex strings in the form `"#rrggbb"`. Hint: write a helper function to pad hex values with a leading 0 if they are only one digit.

Index

Milton Keynes UK
Ingram Content Group UK Ltd.
UKHW031131141024
449569UK00006B/269

9 781138 082557